Hugh Dalziel

The Greyhound

It's History, Points, Breeding, Rearing, Training, and Running

Hugh Dalziel

The Greyhound

It's History, Points, Breeding, Rearing, Training, and Running

ISBN/EAN: 9783743372955

Manufactured in Europe, USA, Canada, Australia, Japa

Cover: Foto ©berggeist007 / pixelio.de

Manufactured and distributed by brebook publishing software (www.brebook.com)

Hugh Dalziel

The Greyhound

MONOGRAPHS ON BRITISH DOGS.

THE GREYHOUND;

ITS

HISTORY, POINTS, BREEDING, REARING,
TRAINING, AND RUNNING.

By HUGH DALZIEL,

Author of "*British Dogs*," "*The Diseases of Dogs*," &c., and Kennel
Editor of "*The Bazaar*."

ILLUSTRATED.

LONDON:
L. UPCOTT GILL, 170, STRAND, W.C.

LONDON :
A. BRADLEY, LONDON AND COUNTY PRINTING WORKS,
DRURY LANE, W.C.

Preface.

This Monograph on the Greyhound is one of a series, the object of which is to meet the requirements of the very large and increasing class of fanciers who take special interest in one or other variety of the dog.

Each Monograph is an enlargement from "British Dogs," giving such additional matter as the modern fancier requires, but which is outside the scheme of the comprehensive work, written for the lover of the dog in general, who does not seek for the completeness of detail which the specialist demands.

The cultivation of the dog in general, and especially in the direction of a clearer separation of distinct varieties, has been greatly stimulated by our competitive exhibitions, and the consequent increased demand, at home and abroad, for pure bred specimens. This is strongly exemplified in the large and increasing number of special clubs, each devoted to the interests of one particular breed, and which authoritatively defines the standard of excellence by which it shall be judged. These circumstances seemed to demand the present series of Monographs, which are designed to serve the particular interest of each fancier.

It will be found that the club standard—where such exists—has been given due weight and prominence to; and as I have, in many instances, received the cordial and much valued assistance of gentlemen accepted as high authorities on their special breeds, I hope the result of these co-operative efforts will prove acceptable and useful to those for whom they have been made.

HUGH DALZIEL.

Contents.

	PAGE
Introduction	1
The Modern Greyhound	9
Coursing	37
Breeding	51
Rearing	63
Training	70
Waterloo Cup Winners	82

THE GREYHOUND.

INTRODUCTION.

Classification according to General Conformation—Points in which Dogs of the Greyhound Group agree—Probable Origin—The Smooth Greyhound—Introduction of the Greyhound to Britain—Arrian—Holinshed—Elfric Duke of Mercia's Kennels—Strutt's "Sports and Pastimes"—Ossian's Description of a Royal Hunting Call—The Deerhound and Irish Wolfhound—The Celtic Greyhound—Edward the Confessor and his Hounds—King John a Lover of Hunting and of the Greyhound—Gelert—Edmund de Langley's "Mayster of Game"—"Booke of St. Alban's"—Coursing for Wagers—Caius' "Englishe Dogges"—The Gazehound—Queen Elizabeth and Coursing—The Duke of Norfolk's Code of Laws of the Leash—Common Source of Wolfhound, Deerhound, and Greyhound.

ANYONE who takes a comprehensive view of the varieties of dogs exhibited at any of our largest shows will find it easy, and indeed natural, to group them, independent of catalogue classifications, according to the general formation and prominent features in which they agree with each other, although differing in minor points, such as coat and colour.

The Greyhound, Deerhound, Irish Wolfhound, Whippet, and even the mongrel Lurcher, show a conformation in common; and extending observation to the classes for foreign dogs, we find the graceful Persian, the Siberian, Circassian, and Pyrenean Wolfhounds, and also the little Italian Greyhound, as well as occasional visitors from the far East, included in this elastic classification by right of the general lines on which they are built.

We have in all, although not equally developed, the same elongated head, long and flexible neck, deep chest, tucked up or girt loins, and the sweeping quarters, which taken together indicate capacity for great speed.

In Europe, Asia, and Africa, we meet with dogs no one would hesitate to class with those named above; but none of the dogs of America, so far as I know, approach them very closely.

The Greyhound group stands out very boldly from all others; and whether all its varieties came originally from the same stock (some *Canis primœvus*, as Darwin suggests), or a species of wolf existent or extinct, it has for ages been recognised as constituting a very distinct type, and from remote history has formed, as it still does, a very important section of our British dogs.

It is probable that the whole of those I have enumerated above which belong to this country, and probably some of the foreign ones, are from the same stock, modified by selection and occasional crosses; as, for instance, the large muscular Irish Wolfhound may have had a strain of the old fierce war dog used by the Celtic natives of these Isles; but it is certain, that at a very early period the wide distinction which separates our quick-footed Hounds from the slower Hounds working by scent, and the still more powerful *Canis pugnaces* of the Molossian or Mastiff type, was recognised as clearly as we now separate them from our Hounds, Spaniels, and Mastiffs.

The smooth Greyhound is now by far the most important, as he is the most perfect, representative of the group, showing in the greatest perfection the qualities for which the whole have always been distinguished and valued.

It is generally believed that the Greyhound was brought into these Islands by a Celtic tribe who spread over Ireland and the Western islands and mainland of Scotland, according to Holinshed during the third century, but much earlier in the opinion of others, who hold that the great Celtic wave spread over Europe, reaching these Islands B.C. 500; but it is admittedly impossible to definitely fix dates. Arrian, writing about the third decade of the second century, gives a full and accurate description of the dog, and calls it a Celtic Hound. Holinshed, in his description of Ireland, says: "The Greihounde of King Cranthylinth's dayes was not fetched so far as out of Grecia, but rather bred in Scotland." (Cranthylinth, or Crathilinthus, was the eighteenth of the kings of Scotland, and began to reign in the year 277.)

Evidence exists that the Greyhound formed part of the kennels of Elfric, Duke of Mercia, for in some curious dialogues written by him in Latin, and translated by Turner, the following occurs: "'I am a hunter to one of the kings.'—'How do you exercise your art?' 'I spread my nets, and set them in a fit place, and instruct my Hounds to pursue the wild deer till they come to the nets unexpectedly, and so are entangled, and I slay them in the nets.'—'Cannot you hunt without nets?' 'Yes, with *swift Hounds* I follow the wild deer.'—'What wild deer do you chiefly take?' 'Harts, boars, and reindeer and goats, and sometimes hares.'" In the Cotton Library, also, there exists a manuscript of the ninth century in which a Saxon chieftain and his huntsman, with a brace of Greyhounds, are portrayed. This picture is copied by Strutt in his "Sports and Pastimes," and shows a couple of dogs with something of the type of head, but shorter in body and tail than, our Greyhounds, about to be slipped at wild swine. I am bound to say that the figures of these dogs as reproduced by Strutt, so far as they can be relied on to represent a breed, are more like the Great Dane of our shows in head and carriage of stern (which latter is Houndlike); but the back is shorter, and the ears appear short, pointed, and erect, as if cropped. It is the more important to notice this, as we have the assurance of Strutt that the engravings are faithful copies of ancient ones. Far more Greyhound-like is the dog represented in the picture, "The Unearthing of a Fox," from a manuscript of the fourteenth century in the Royal Library.

Among the wild clans of the North the ancestors of our Deerhounds were cherished, and used by those hardy hunters in the pursuit of the stag, as well as in the destruction of the wolf; and the stealing of one that excelled in size, swiftness, and courage, by a clan that had been the guests of another at a hunting party, led to a furious and bloody combat. And however apocryphal the songs of Ossian may be, the writer touches a genuine chord in the national sympathies in singing the praises of the dogs of Fingal in his description of a royal hunting "call." Said Fingal: "Call my dogs, the long-bounding sons of the chace. Call white-breasted Bran, and the surly strength of Luath. Fillan and Fergus! blow my horn, that the joy of the chace may arise; that the deer of Cromla may hear and start at the Lake of Roes. The shrill sound spreads along the wood. The sons of heathy Cromla arise. A thousand dogs fly off at once, gray-bounding through the heath. A deer fell by every dog, and three by the white-breasted Bran."

That the Deerhound and Irish Wolfhound were, if not identical, very closely allied, I think there can be no doubt; and with such game as they were fitted to cope with in abundance, the fugacious hare was thought little of; but in the lower and more open countries the lighter built and more nimble dogs would be used for that quarry. Arrian, describing the Celtic Greyhound, refers to both smooth and long-haired; and I think it in the highest degree probable that all are from the same stock, for we know that quantity and quality of coat readily change, and, according to domestic treatment, quite alter in character in a few generations, whilst variation in colour is the common inheritance of domestication.

According to William of Malmesbury, Edward the Confessor "took great delight to follow a pack of swift hounds in pursuit of game, and to cheer them with his voice;" these were, probably, hounds running by scent. But the same writer, enumerating the dogs of the chase, includes Greyhounds as favourite dogs with the sportsmen of that time.

King John was a lover of hunting and of the Greyhound, and the gallant Gelert, made famous in Spenser's poem, was said to be a gift of this king to his son-in-law Llewellyn. That the story must be admitted to be mythical does not altogether destroy its value. John was at heavy charges in the maintenance of his kennels, including Wolfhounds and Greyhounds, and his son and successor Henry III., who instituted the severest of Forest Laws, kept up the sport of hunting, in which these and other varieties of dogs were used.

Edmund de Langley, fourth son of Edward III., who was born A.D. 1341, became Master of Hounds and Hawks to Henry IV., and wrote a Treatise called "Mayster of Game," for the pleasure and instruction of Prince Henry — afterwards Henry V. — in which the Greyhound is minutely described.

Following shortly after De Langley we have the celebrated "Booke of St. Alban's,"* by Dame Juliana Berners, or Barnes, published by Winkin de Worde, 1486, in which Greyhounds and hare-hunting, as well as stag-hunting, are referred to and explained.

In the time of Henry VIII. it was a boast of manhood on the part of the young gallants, among other accomplishments—

> . . . . . to nourishe up and fede
> The Greyhounde to the course.

During this reign we have the first mention of coursing for wagers. Jesse quotes from the accounts of expenditure

---

* A fac-simile reprint of the "Booke of St. Alban's," containing the Treatises on Hawking, Hunting, and Heraldry, has been issued by Mr. Elliot Stock, 62, Paternoster Row, London, E.C.

of the King: "Sir William Pykering received forty-five shillings for a course that he won of the king's grase in Eltham Park against his dog; and another person twenty-two and sixpence for bets that he won of the king in Eltham Park. Also the Lord Rochford—forty-five shillings for a wager he won with a brace of Greyhounds at Mote Park."

In the reign of Elizabeth, Dr. Johannes Caius wrote his "Englishe Dogges,"* in which the Greyhound is described, but not with that accuracy of detail which we find in the writings of Gervase Markham, who followed a generation later, or of Edmund de Langley, who preceded him by about two centuries.

Caius recognises the distinction in size, coat, and the purposes to which the dogs were put which answer to our Deerhounds and Greyhounds of to-day. "Some," he remarks, "are of a greater sorte, and some of a lesser; some are smooth skynned, and some are curled; the bigger, therefore, are appoynted to hunt the bigger beasts, and the smaller serve to hunte the lesser accordingly."

The Gazehound, one of this group, used to single out and pursue the wounded or selected deer by sight alone, is also mentioned by Caius, but by him wrongly termed "*Agaseus*," which really represents the Beagle, and had been applied to that dog for centuries previous to his time.

In Elizabeth's reign the diversion of coursing became more fashionable than it had ever been previously, and Her Majesty personally enjoyed the sport of coursing stags with Greyhounds. It is recorded that, on a visit to

---

* A verbatim reprint of this book has been published by L. Upcott Gill, 170, Strand, London.

Cowdry Park, the seat of Lord Montacute, the Queen saw sixteen bucks pulled down by Greyhounds after dinner, the bucks having fair law. These dogs were probably of the strong Deerhound type.

Hare-coursing now became established on a firmer and better basis than it had occupied, owing to the formation of the laws of the leash into a regular code by the Duke of Norfolk, which, with alterations not affecting their principle, rules at the present day. The development of coursing from that date to the present time I shall treat of more fully hereafter.

It is impossible to trace the divergence from the original Greyhound, and the modifications which have resulted in the varieties we now possess; but all history and records of sport seem to point to the fact that Wolfhound, Deerhound, and Greyhound, sprang from the same source, although, probably, not each alike bred in purity. In the case of the latter the type has, by careful breeding, become fixed, and even in the most trivial features there is a closer likeness between the individuals of the race than ever before existed. Each variety of the group will be found dealt with separately, and at length, in "British Dogs;" in this book I confine myself to the modern Greyhound. But the following description, said to be from an old MS., though of what date I do not know, applies to the generic character of the purer varieties of the group:

> The Greyhound, the great Hound, the graceful of limb,
> Rough fellow, tall fellow, swift fellow, and slim;
> Let them sound o'er the earth, let them sail o'er the sea,
> They will light on none other more ancient than he.

THE MODERN GREYHOUND.

> The Slowhound wakes the fox's lair,
> The Greyhound presses on the hare.
> —ROKEBY, Chap. iii.

The Modern Greyhound: Elegance of; Fitness for Designed Purpose—Derivation of the Word Greyhound: Various Opinions on—Gesner—The Grew—Modifications in the Greyhound — Restrictions as to Keeping Greyhounds—Coursing: Popularity of; Antiquity of—Greyhounds used for Hunting the Wolf, Boar, Deer, &c.—Distinguishing Characteristics of Greyhound Group—Essential Points in a Greyhound—Sir Walter Scott—Old Prejudices against Coursing — Somerville — Mistakes of the Tyro as to Coursing—Slipping the Dogs—The Run up—A Kill of Merit—Ovid's Description of a Course—Judging Greyhounds—Dame Juliana Berners' "Properties of a Good Grehounde"—Detailed Description of Properties: Head, Jaws, Teeth, Eyes, Ears, Neck, Back, Chest, Loins, Tail, Colour, Size—Crossing with the Bulldog—Summary of Points of Modern Greyhound—Greyhounds and the Show Bench—Measurements of Waterloo Cup Winners and Show Dogs.

THE particular variety of *Canes venatici grayii* of which I will now treat, and which possesses an inherent right to occupy the highest place in the group of dogs hunting by keenness of sight and fleetness of foot, is the modern

British Greyhound. I say British, for the time has gone by when we could speak of English, Scotch, or Irish Greyhounds in any other than the past tense; and the modern Greyhound—the most elegant of the canine race, the highest achievement of man's skill in manipulating the plastic nature of the dog, and forming it to his special requirements—as he is stripped, in all his beauty of outline and wonderful development, not only of muscle, but of that hidden fire which gives dash, energy and daring, stands revealed a manufactured article, the acme of perfection in beauty of outline and fitness of purpose; and whether we see him trying conclusions on the meadows of Lurgan, the rough hillsides of Crawford John, or for the blue ribbon of the leash on the flats of Altcar, he is still the same—the dog in whom the genius of man has so mingled the blood of all the best varieties of the Celtic *Canes celeres* that no one can lay special claim to him. He is a combination of art and Nature that challenges the world, unequalled in speed, spirit and perseverance, and in elegance and beauty of form as far removed from many of his clumsy ancestors as an English thoroughbred from a coarse dray-horse.

It is not my intention, further than I have already done, to attempt to trace the history of the Greyhound, or to follow his development from the comparatively coarse but more powerful dog from which he probably derives his origin. It is clear, however, that our ancestors had two thousand years ago developed his form, swiftness, and wind, so as to enable him to run down the deer and hare at speed. The very name has long been a bone of contention among etymologists; but however interesting to the scholar, the discussion possesses few attractions for the general reader, the ingenious guessing and nice hair-splitting proving often more confusing than profitable. Not to pass the subject over in complete silence, I may observe, that whilst some contend that the name *Canis Græcus* points to a Greek origin, others derive the name from "grey," *gre*, or *grie*, supposed to be originally the prevailing colours, others, with apparently greater reason, suppose the name to have been given on account of the high rank or degree the dog held among his fellows. The latter is the meaning attached to it by Caius, who says: "The Grehound hath his name of this word *gre*, which word soundeth *gradus* in Latine, in Englishe *degree*, because, among all dogges these are the most principall, occupying the chiefest place, and being simply and absolutely the best of the gentle kinde of Houndes." Dansey, the translator of Arrian, thinks this fanciful, but at the same time points out that the word *gre* is used for degree by Gawin Douglas, Thomas the Rhymer the Prophet of Ercildoun, the author of the metrical romance "Morte Arthur," and by Sir David Lindsay in his satirical poem, "The Complaint of the King's Auld Hound, Bagsche," when addressing the King's new favourite Houndes: "Though ye stand in the highest gre"; and he (Dansey) acknowledges: "Whimsical, therefore, as Caius's tracing of the term may be, we cannot view it as utterly untenable."

In addition to this, I may point out that the word, bearing the same signification, is to be found in Burns and Scott, and that there is a common Scotch phrase "to bear the gree," meaning to be decidedly victor, or to take precedence of others.

The translator of Arrian "would rather seek the origin of the English name in the prevailing colour of the dog," in which the numerous commixtures of colours forming the various shades of grey predominate. "Stonehenge" leans to this view; and he also points out that "no other breed has, I believe, the blue or grey colour prevalent." The blue colour is, however, common to the Great Dane.

Again, it has been sought to trace the name to the grey, the old name for the badger—hence Grey-hunde, or Badger-hound; but although the suggestion comes from so high an authority as Skinner, I cannot but think it entirely fanciful, for a dog of the Greyhound type is not needed for such slow game as the badger.

Gesner spells the name Grewhound; and Golding in his translation of Ovid uses the contraction of that compound, Grewnd, which is common enough:

And even as when the greedy Grewnde doth course the sillie haro,
Amiddes the plaine and champion fielde, without all covert bare.

In the south of Scotland the Greyhound is still commonly called the Grew; and in "Guy Mannering," Dandie Dinmont includes in the list of his dogs, "six Terriers, twa couple of Slowhounds, five *Grews*, and a wheen ither dowgs."

The Greyhound having been always kept for the chase, would naturally undergo modifications with the changes in the manner of hunting, the nature of the wild animals he was trained to hunt, and the characteristics of the country in which he was used; and having always, until very recent times, been restricted to the possession of persons of the higher ranks, he would have greater care, and his improvement be the better secured. That his possession

THE MODERN GREYHOUND. 13

was so restricted is shown by the Forest Laws of King Canute, which prohibited anyone under the degree of a gentleman from keeping a Greyhound; and an old Welsh proverb says: "You may know a gentleman by his horse, his hawk, and his Greyhound." In the Welsh laws of Howel Dda (who died 948), the King's Buckhound, or Covert-hound, is valued at a pound, and his Greyhound at six-score pence. In the Code of 1080, and the Dimetian Code of 1180, the Greyhound is valued at half that of the Buckhound.

The alteration in the game laws of modern times, coupled with the great increase of wealth and leisure, have, by giving impetus to the natural desire for field sports characteristic of Englishmen, led to the present great and increasing popularity of coursing, and consequent diffusion of Greyhounds through all classes, heightening an honourable competition, and securing a continued, if not a greater, care, and certainty of the dog's still further improvement.

It is impossible to separate the Greyhound from coursing as we understand it; for although the sport existed, and was practised in a manner similar to our present system, some seventeen hundred years ago, as described by Arrian in the second century, the thorough organisation of the sport and the condensation of the laws governing it are not only essentially British, but, in their present shape, quite modern; and it is the conditions of the sport that have produced the Greyhound of the day, to which the words—

> They are as swift as breathed stags,
> Aye, fleeter than the roe,

are more applicable than to any of its predecessors.

If we go back to the earlier centuries of the history of our country, we find the Greyhound used in pursuit of the wolf, boar, deer, &c., in conjunction with other dogs of more powerful build. Still, we can easily perceive that, to take a share in such sports at all, he must have been probably larger, certainly stronger, coarser, and more inured to hardships, whilst he would not be kept so strictly to sight hunting as the demands of the present require; but the material out of which the present dog has been made was there, and his form and characteristics, even to minute detail, were recognised, and have been described with an accuracy which no other breed of dogs has had the advantage of, else might we be in a better position to understand the value of claims for old descent set up for so many varieties. And to these descriptions I propose to refer, to endorse, as well as to make still more clear and emphatic, the points of excellence recognised as correct by modern followers of the leash.

The whole group to which the Greyhound belongs is distinguished by the elongated head; the parietal, side and upper, or partition bones of the head, shelving in towards each other; high, proportionate stature, deep chest, arched loins, tucked-up flank, and long, fine tail; and such general form as is outlined in this description is seen in perfection in the Greyhound. To some it may sound contradictory to speak in one sentence of elegance and beauty of form, and in the next of a tucked-up flank; and Fox Terrier and Mastiff men, who want their favourites well-ribbed back, with deep loin, and flanks well filled, to make a form as square as a prize Shorthorn, may object; but we must remember, that beauty largely consists in fitness and aptitude for the uses designed and the position to be filled.

This being so, in estimating the Greyhound's claim to be the handsomest of the canine race we must remember for what his various excellencies, resulting in a whole which is so strikingly elegant, are designed. Speed is the first and greatest quality a dog of this breed can possess; to make a perfect dog there are other attributes he must not be deficient in, but wanting in pace he can never hope to excel. The most superficial knowledge of coursing or coursing literature will show this, and it is a quality which, although developed to its present high pitch, has always been recognised as most important. Chaucer says:

> Greihounds he hadde as swift as fowl of flight;

And again—following the example of the immortal scoundrel, Wegg—to drop into poetry, Sir Walter Scott in his introduction to "Marmion" thus eulogises the speed of the Greyhound:

> Remember'st thou my Greyhounds true?
> O'er holt or hill there never flew,
> From leash or slip there never sprang,
> More fleet of foot, more sure of fang.

Well does he deserve the encomium of Markham, who declares he is "of all dogs whatsoever the most princely, strong, nimble, swift, and valient."

In addition to speed, the dog must have strength to last out a severe course, nimbleness in turning, the capacity to catch and bear the hare in his stride, good killing powers, and vital force to give him dash, staunchness, and endurance. What a dog possessing these qualities should be like I shall, by the assistance of the keenest and most experienced observers and writers on the subject, endeavour to show; and whilst gladly sitting at the feet of modern Gamaliels, will not slight the wisdom of the past, but will offer gleanings from the works of old that may prove both interesting and instructive to the tyro, although as a tale that hath been told to many; and in defence of such a course let me quote Geoffrey Chaucer:

> For out of the old fieldis, as men saith,
> Cometh all this new corn from year to year;
> And out of olde bookis in good faith,
> Cometh all this new science that men lere.

It will be unnecessary at this point to enter on any lengthened dissertation on coursing as at present practised, and it will suffice to say that the sport has, in fact, lived down the old prejudices which existed against it, as expressed in the words of Somerville, whose tastes preferred

> The musical confusion
> Of Hounds and echo in conjunction;

and who, with unjust prejudice, penned an undeserved censure against followers of the leash when he wrote:

> A different Hound for every different chase
> Select with judgment; nor the poor timorous hare,
> O'er matched, destroy; but leave that vile offence
> To the mean, murderous, coursing crew.

I have no doubt, however, that Somerville, who was a thorough sportsman, had in his mind when penning these lines the poacher and currant-jelly courser, who do not hesitate at the means, or how they o'ermatch the hare, so

THE MODERN GREYHOUND. 17

that their supreme object of filling the bag is attained; for with such a murderous crew the hare gets no law.

Before discussing the points of the dog *seriatim*, it will, I think, be necessary to briefly glance at what a dog is required to do in a course, and that for two reasons: First, because I hold that all dogs should be judged in the show ring by their apparent suitability for their special work; and secondly, because this book may fall into the hands of many who are real lovers of the dog, and genuine sportsmen at heart, but who, from various circumstances, have never had an opportunity of seeing a course, or that so rarely as to be practically unacquainted with its merits.

The remarks of the inexperienced on a course are often amusing. The most common mistake made by the tyro is that the dog which kills the hare always wins, irrespective of other considerations—a most excusable error on the part of the novice, as in most or all other descriptions of racing the first at the post or object is the winner; but in coursing it is not which is first there, but which has done most towards accomplishing the death of the hare, or put her to the greatest straits to escape. Be it here understood, that the object of the courser and that of the dogs differ materially. The dogs' object is the death of the hare; the courser's object is to test the relative speed, working abilities, and endurance of the competitors, as shown in their endeavours to accomplish their object: and the possession of the hare is of little consequence, except to the pothunter or currant-jelly devotee, who is quite out of the pale of genuine coursing society.

C

Although what I am going to say will be as stale and tiresome to—and as likely to create a smile in—many as listening to a child's first lesson in the alphabet, I consider it, for the reasons already given, necessary. Two dogs only are slipped at a hare—and this has always been the honourable practice in this country. Even the Greek courser Arrian recognises this, saying: "Whoever courses with Greyhounds should neither slip them near the hare, nor more than a brace at a time"; and in Turberville's "Observations on Coursing" we find the maxim: "If the Greyhounds be but yonge or slow you may course with a lease at one hare, but that is seldom seen, and a brase of dogges is ynow for such a poore beaste."

The hare being found or so-ho'd, and given law—a fair start of, at one time, more than a hundred yards, but now reduced—the dogs are slipped. In the run up, as in after stretches following a turn, the relative speed of the dogs is seen; but the hare, being pressed, will jerk, turn, and wind in the most nimble manner, testing the dogs' smartness in working, suppleness, and agility in making quick turns, and "it is a gallant sport to see how the hare will turn and wind to save herself out of the dogge's mouth, so that, sometimes, when you think that your Greyhound doth, as it were, gape to take her, she will turn and cast them a good way behinde her, and so save herself by turning, wrenching, and winding." It is by the practice of these clever wiles and shifts that the hare endeavours to reach her covert, and in closely following her scut, and o'ermastering her in her own devices, that a Greyhound displays the mastery of this branch of his business, in which particular a slower dog will often excel an opponent that has the foot of him in the stretches; but with this working power, a facility in making short turns, speed must be combined, or it stands to reason points could not be made, except on a comparatively weak hare. It is, therefore, important that the conformation of the dog should be such as to combine speed with a strength and suppleness that will, as far as possible, enable him to control and guide the velocity with which he is moving, as his quick eye sees the game swerve or turn to one side or another.

As the death of the hare when it is a kill of merit—that is, when accomplished by superior speed and cleverness, and not by the accident of the foremost dog turning the hare, as it were, into the killer's mouth—is a consideration in reckoning up the total of good points made, it is important that the dog should be formed to do this, picking up and bearing the hare in his stride, and not stopping to worry her as a terrier would a rat; and here many points come in which should be narrowly scanned and compared in the show ring, but too often are not, and these I will allude to in going over the several points.

In addition, there are other requirements for which the dog must possess qualities to make him successful in the field and give him a right to a prize in the show ring, and which will be noticed in detail. A good idea of a course, with the gallant efforts of pursuer and pursued, is given in the following lines from Ovid, translated by Dryden:

> As when the impatient Greyhound, slipped from far,
> Bounds o'er the glade to course the fearful hare,
> She in her speed does all her safety lie,
> And he with double speed pursues his prey,
> O'erruns her at the sitting turn; but licks
> His chaps in vain; yet blows upon the flix.
> She seeks the shelter which the neighbouring covert gives,
> And, gaining it, she doubts if yet she lives.

The laws and details of coursing, and the preparation of the dogs, are fully treated of hereafter, but the preceding remarks appeared necessary to a clear comprehension of the description we shall now enter upon.

In forming an opinion of a dog—whether in selecting him for some special purpose of work, or merely choosing the best out of a lot in the prize ring—first impressions, which are occasionally deceptive, get confirmed into prejudices and mislead the judgment. But in the great majority of cases, to the man who knows what he is *looking at*, what he is *looking for*, and what he has a reasonable right to expect, the first impression conveyed to the mind by the general outline or contour, and the way it is filled in, will be confirmed on a close, critical, and analytical examination of the animal point by point; and it is only by such close and minute examination that a judge can become thoroughly master of his subject, and arrive at a position where he can give strong, clear, and intelligible reasons for the opinions he has formed and the decision he has given. Moreover, there is that to be weighed and taken into account in the final judgment on the dog's merits which is referable to no part alone, and which can only be appreciated on taking him as a whole—that is, *life*—that indefinable something which evades the dissector's knife, yet permeates the whole body; the centre power, which

THE MODERN GREYHOUND. 21

is the source of movement in every quivering muscle, and is variously seen in every action of the dog, and in every changing emotion of which he is capable. This I conceive to be the only difficulty in the way of judging by points, and it is not insuperable: this is, probably, what is often meant by *condition* and *quality*.

The judge must, however, as already said, consider, and if need be describe, not only the general appearance of the animal and the impression he conveys to his (the judge's) mind, but, as it were, take him to pieces, assessing the value of each particular part according to its fitness for the performance of the special function for which it is designed, and under the peculiar conditions in which it will have to act; and having done so, he will find his first opinion confirmed precisely in the ratio of his fitness to judge.

Before taking the points one by one, I must give the description of a Greyhound as laid down in the doggerel rhymes of the illustrious authoress of "The Booke of St. Alban's," Dame Juliana Berners, or Barnes, somewhile Abbess of Sopewell, and since described as "a second Minerva in her studies and another Diana in her diversions." It would be sheer heresy to write of Greyhounds without introducing her description, so universally has this been done; I therefore give it in full, which I have never seen done by any modern authority. In doing so, I must confess there are two lines that to me are somewhat obscure. I, however, venture to suggest that in his eighth year he is only a *lick ladle*—fit to lick a trencher; and in his ninth year, cart and saddle may be used to take him to the tanner.

The Properties of a Good Grehounde.

A Grehound shold be heeded lyke a snake
And neckyd lyke a drake,
Footed lyke a catte,
Tayllyd lyke a ratte,
Syded lyke a teme,
And chynyd lyke a beme.
The fyrst yere he must lerne to fede,
The second yere to felde him lede,
The thyrde yere he is felowe lyke,
The fourth yere there is none syke,
The fyfth yeare he is good enough,
The syxte yere he shall hold the plough,
The seventh yere he woll avaylle
Grete bytches for to assaylle,
The eygthe yere licke ladyll,
The nynthe yere cartsadyll;
And when he is comyn to that yere
Have him to the tannere,
For the best Hounde that ever bytche had
At nynthe yere he is full badde.

To begin the detailed description with *the head*—which includes jaws, teeth, eyes, ears, and brain development—first the general form must be considered. It must be quite evident that "headed like a snake" cannot mean "like a snake's head," which is short, flat, and blunt, or truncated. I understand the Abbess to use the snake itself, not its head only, as a simile of the length and thinness of the Greyhound's head.

Arrian says: "Your Greyhounds should have light and well-articulated heads, whether hooked or flat-nosed is not of much consequence, nor does it greatly matter whether the parts beneath the forehead be protuberant with muscle. They are alone bad which are heavy-headed, having thick nostrils, with a blunt instead of a pointed termination." Edmund de Langley, in his "Mayster of Game," says:

"The Greihound should have a long hede and somedele grete, ymakyd in the manner of a luce; a good large mouth and good sessours, the one again the other, so that the nether jaws passe not them above, ne that thei above passe not him by neither"; and coming down to Gervase Markham, in the sixteenth century, we have his description: "He should have a fine long leane head, with a sharp nose, rush grown from the eyes downward."

The general form and character of the head is here pretty fairly sketched, and we see a very close agreement between these old authorities. It appears to me that the "Mayster of Game" was the most happy in his illustration, "Made in the manner of a luce"—that is, a full-grown pike—as the heads of the Greyhound and pike will bear a fair comparison without straining; and who can say it was not the exigencies of rhyme that compelled our sporting Abbess to set up for us that stumbling-block, the head of a snake? No doubt she thought of the excellent illustration the neck of the drake offered her, and had to find a rhyme to it; but she might with as great propriety have written:

The Grehound should be heeded like a luce
And neckyd like a goose.

The force of illustration lost in the second line is more than compensated for by the strength of the first. Markham is right in desiring a "long leane head," though even that may be carried to a fault: but we do not want the "part beneath the forehead protuberant of muscle"; and the "heavy-headed, with thick nostrils and a blunt nose," I must, with Arrian, discard altogether as thoroughly bad, too slow, and certain to be "too clever by half." Looking at the whole head, we see by the sloping in of the side walls of the skull how the brain capacity is diminished, and how the elongation and narrowing of head and jaws have almost obliterated the olfactory organs, the internal cavities becoming contracted, and presenting so much less surface that the scenting powers are necessarily limited, although it is a mistake to suppose that they are entirely lost. This is just what we want in the Greyhound; he must run by sight, never using his nose; he must have the brain developed where it shows courage, not intelligence. When a Retriever has to puzzle out a lost bird his nose and his intelligence are both put to the test, and the higher the development the better the dog. And as we find the intellectual faculties highest in those dogs with most brain, so we select our Retrievers thus formed; but as this would be a disadvantage in the Greyhound, which we want to run honest and fair—such as Justice Shallow in the "Merry Wives of Windsor" describes:

> He is a *good* dog and a *fair* dog;
> Can there be more said?—he is *good* and *fair*—

we select them without this intellectual development, by use of which they would soon study the wiles and shifts of "poor Wat," and, to save their wind and legs, "run cunning"—that is, do a "waiting race," the cunning dog allowing his fellow to do the work, whilst he hangs back for the hare to be turned into his mouth. A Greyhound should measure well round the head by the ears, which is a sure indication of the courage that gives dash and persistence to his efforts.

By "hooked nose" I presume Arrian to mean that the upper jaw protrudes; but that would decidedly be a fault, as a dog so formed would be at a disadvantage in holding and killing his hare. This formation, called overshot, or pig-jawed, is often met with in various breeds of dogs, but if at all excessive is most objectionable. The opposite to that is sometimes seen, and we have them undershot, though such cases are comparatively rare, and owe their origin to the cross with the Bulldog, which has been resorted to to give stamina, courage, and staunchness. The form to be desired is the level mouth, with the "good sessours, the one again the other."

There is a formation of muzzle met with which is slightly ridged or Roman nosed; if not excessive, this is no detriment to the dog's practical usefulness, although as a matter of taste it may not be considered as adding to the beauty of his appearance. This peculiarity may exist with a good level mouth.

The teeth themselves are important, and should be large, strong, and white, the fangs sharp and powerful—the upper ones just overlapping those in the lower jaw; this is not only necessary for their work, but is always a sign of health.

"*The eye,*" Arrian says, "should be large, upraised, clear, and strikingly bright. The best look fiery, and flash like lightning, resembling those of leopards, lions, or lynxes." Markham says: "A full, clear eye, with long eyelids." The latter peculiarity I have never observed, probably from want of a close attention to the point; but the clear, bright, and fiery eye is always a necessity, although, of course, the condition of the dog and the circumstances under which he is seen must be considered in judging of it; the colour varies with that of the coat, as in all breeds. I have only met with one instance of wall-eye, or china-eye, such as is so common in the peculiar dappled grizzle-coloured Colleys, and that was in a pure bred and very handsome dog with light blue brindle markings; it is certainly a disfigurement, but in no way interferes with the dog's vision.

Of *the ears* Arrian writes: "They should be large and soft, so as to appear broken; but it is no bad indication if they appear erect, provided they are not small and stiff." This description would not be accepted as satisfactory now, as ears are preferred small and free from all coarseness. Neither does Markham's "a sharp ear, short, and close-falling," quite convey the modern idea of a Greyhound's ear; it should be soft, fine in leather, and folded, with the shoulder of the ear strong enough to carry the whole up when the dog is excited or his attention fixed.

The neck is the next point, and it is one of very great importance; it must be long, strong, well-clothed with muscle, yet withal light, airy, and possessing wonderful flexibility and suppleness. Arrian says: "The neck should be long, round, and flexible, so that if you forcibly draw the dogs backwards by their collars it may seem to be broken from its flexibility and softness." The neck is certainly wonderfully pliant, and readily bent to either side at will. Our Royal writer says: "The neck should be grete and longe, and bowed as a swanne's neck"; Markham: "A long neck, a little bending, with a loose, hanging wezand." The last point is not correct, and might convey the idea that there was a looseness of skin underneath: the windpipe, although easily felt, does not hang loose, the whole neck being neat, round, clean made, and elegantly carried. A long neck, as well as a long head, is necessary to enable the dog to pick up, carry, or bear the hare without stopping, which he will do throwing his head up with the hare in his mouth; but a dog with a short neck would have to stoop so in catching his hare, that there would be a very great chance of his coming a "cropper," the force at which he was going throwing him heels over head.

Continuing from the neck we have the broad, square, beam-like *back*, of good length and great strength; without this the dog could not endure the exhaustive process of the "pumpers" he is submitted to. *The chest*, too, must be deep, and fairly wide. Arrian says: " Broad chests are better than narrow; shoulders wide apart, not tied together, but as loose and free as possible; legs round, straight, and well-jointed; sides strong; loins broad, firm, not fleshy, but sinewy; upper flanks loose and supple; hips wide asunder; lower flanks hollow; tail long, fine, and supple; haunches sweeping, and fine to the touch." In respect to the chest, it is needless to say how all-important it is that it should be capacious; but we must get capacity from the depth and squareness, not from the bulged-out, barrel form, which would produce slow movement and a heavy-fronted dog that would soon tire. Take Markham's description in "The Country Farm": "A long, broad, and square beam back, with high, round fillets; he must be deep, swine-sided, with hollow bended ribs and a full brest."

"The Mayster of Game" gives an excellent description: " Her shuldres as a roebuck; the for leggs streght and grete ynow, and nought to hind legges; the feet straught and round as a catte, and great cleas; the boones and the joyntes of the cheyne grete and hard as the chyne of an hert; the thighs great and squarred as an hare; the houghs streight, and not crompyng as of an oxe." The shoulders should be set on as obliquely as possible, to enable the dog to throw his fore legs well forward in his gallop, the shoulder blades sloping in towards each other as they rise; they should be well clothed with muscle, but not fleshy and coarse so as to look loaded; the shoulders should not be tied together, but have plenty of freedom—this, with the strong muscles of the loin, enables the dog to turn fast and cleverly; the elbows must be neither turned out nor in; the bone of the leg must be strong; there must be good length of arm; and the leg below the knee must be short and very strong, and the foot round and cat-like; well sprung knuckles, a firm, hard, thick sole, and large, strong nails, are also essential.

The beam-like back is to give the necessary strength; the deep chest is needed with sufficient width to give plenty of room for the lungs and heart to freely perform their functions; width is needed that the necessary room may be got without making the chest so deep as to be in the way and catch against stones, tussocks, and lumps of turf on rough, coarse ground, when the dog is fully stretched in the gallop; the oblique shoulders enable the dog to throw his legs well forward and close together, thus enabling him to cover a lot of ground at each stride, and also, in connection with his long and supple neck, to throw himself through an astonishingly small meuse. The necessity of sufficient bone, big, strong joints, and muscular legs is apparent where such violent exertion is called for, and the round, cat-like foot, is a necessity of speed: no one would have the wheels of a fast-going gig made as broad in the tyre as those of a four-ton waggon. The soles are required hard and tough that they may stand the wear and tear of rough ground and stony lanes, if these have to be travelled over; the strong claws give the dog purchase over the ground.

The *loins* must be strong; a Greyhound weak there might be fast for a spurt, but he would prove merely flashy, being neither able to endure nor yet be good at his turns. When Markham says, " short and strong fillets," he means the loin—the term being used in speaking of the horse—not the fleshy part of the thigh, which the term might seem to indicate. The hips must be wide asunder, and the hind legs straight as regards each other, "not crompyng as of an oxe "—that is, as we now express it, not cow-hocked—but they must be bent or sickle-hocked, and the thighs with immense and well-developed muscle. The same strength of muscular development is needed as in the fore legs, and especially there should be no weakness below the knee. The dog should stand rather wide behind, and higher than before; the slight width gives additional propelling force, and the higher hind quarters additional speed and power in racing up hill, as hares invariably do, if they can, unless there is temptation of a covert near, a fact quaintly expressed in the " Booke of St. Alban's " :

> " Tell me, Maystre," quod the man, " what is the skyll
> Why the Haare wolde so fayne renne against the hill ? "
> Quod the Mayster, " For her legges be shorter before
> Than behind ; that is the skyll thore."

In respect to *the tail*, all agree it should be long and fine. Markham says: "An even growne long rat's tail, round, turning at the lower end leashward, and full set on between the buttocks." The "Mayster of Game" says: "A catte's tayle, making a ring at eend, but not to hie." The tail, no doubt, acts as a rudder, and as such must play an important part in swerving and turning.

Colour in Greyhounds should go for little; for although many persons have a prejudice in favour of a special fancy, experience proves that there are good of all. In the hunting poem by Gratius, as translated by Wase, we are told to—

> Chuse the Greyhound pied with black and white:
> He runs more swift than thought or wingèd flight.

Arrian considered the colour of Greyhounds of no importance; but most old, and also modern, writers have their preferences. As directly opposite to the opinion just quoted, Oppian objects to white and black, as too sensitive to heat and cold. De Langley says: "Of all manere of Greihoundes there byn both good and evel; Natheless the best hewe is rede falow, with a black moselle." "Stonehenge" says the colours preferred are black, and red or fawn, with black muzzle; and it may be worth notice that, in quoting him, "Idstone" falls into the singular mistake of saying they should have red muzzles. Turberville mentions white, fallow, dun, and black as the preferable colours of Hounds; and that the dun is an old favourite colour may be inferred from the following lines from an ancient metrical romance:

> " Sir, yf you be on hunting bounde,
> I shall you gyve a good Greyhounde
> That is dunne as a doo;
> For as I am trewe gentlewoman,
> There was never deer that he at ran
> That myght yscape him fro."

At the sale of the Greyhounds of that eminent courser, Lord Rivers, in May, 1825, a list of which is given in Goodlake's "Courser's Manual," there were, out of fifty-two dogs, twenty-three all black, fourteen all blue, six red, four blue and white, and one all white. There are still many coursers who prefer the pure black or the red; but the following short list, taken from the "Coursing Calendar," shows good Greyhounds of many different colours: Scotland Yet and her sons Canaradzo and Calioja were white; Cerito, fawn and white; Lobelia, brindled and white; Lady Stormont, black and white; Master M'Grath, black and white; Beacon, Blue Light, and Sapphire, all blue; High Idea, blue ticked; Bed of Stone, Bab at the Bowster, and Sea Cove, red; Cauld Kail, red ticked; Mocking Bird, Cashier, and Black Knight, all black; Landgravine and Elsecar, brindled.

As regards *size*, the medium-sized dog is preferred by most. There is a considerable difference, both in height and weight, between the dog and bitch. Prejudice against small dogs received a shock by the double victory of Coomassie in the Waterloo Cup, she being a bitch of only 42lb. running weight; and her appearance, also, was not prepossessing, her colour being a washed-out fawn. Again, Penelope II., the runner up for the Cup in 1886, weighed but 41lb., her victor, Miss Glendyne, weighing 54lb.; this seeming to confirm the courser's adage, "A good big one will always beat a good little one."

A cross with the Bulldog was resorted to by Lord Orford with the object of giving additional *courage* to the Greyhound, and it was held to have produced that result; but subsequent experiments in that way have not, I believe, resulted successfully.

Summary of Points of Modern Greyhound.

The Head.—Long and lean, but wide between ears, measuring in girth, just before or close in behind, about 15in. in a dog 26in. high, with a length from occiput to nose of about 10in. to $10\frac{1}{2}$in.

The Ears.—Set on well back, small and fine in the flap, falling gracefully with a half fold back, exposing the inner surface. Erect or pricked ears are now seldom seen, and are disliked.

The Eye.—Varying in colour; must be bright, clear, and fiery.

The Teeth.—Strong and white, the upper canines, with the slight curve they possess, clipping those of the lower jaw. (Value 15.)

The Neck.—Length and suppleness are of great importance, to enable the dog to seize the hare as he runs at full speed. It is elegantly bent or arched above the windpipe, giving it a slightly protuberant form along the lower surface, the whole gradually swelling out to meet the shoulders. (Value 10.)

Chest and Fore Quarters (including shoulders and fore legs).—*The Chest* must be capacious, and the room obtained more by depth than width, to give free action to the heart and lungs. *The Shoulders.*—The scapula, or shoulder blade, must be oblique, that the fore legs may be readily stretched well forward. The arm from shoulder to elbow, and fore arm from elbow to knee, both of good length, and short from knee to the ground. *The Elbows* must not turn either in or out, but be in a straight line, so that the action may be free. *The Muscles* for expansion and retraction of the several parts of legs and shoulders must be large and well-developed. (Value 20.)

Loin and Back Ribs.—The back should be broad and square, or beam-like, slightly arched, but not approaching to the wheel back of the Italian Toy Greyhound. The loin wide, deep, and strong, the muscles well-developed throughout, so that, although the flank is cut up, it yet measures well round—and this is important, as showing strength. (Value 15.)

Hind Quarters.—Strong, broad across, the stifles well bent; first and second thigh both big with muscle; the legs rather wide apart, and longer than the fore legs, short from the hock to the ground. (Value 20.)

Feet.—Round, with the toes well sprung, the claws strong, and the pad, or sole, compact and hard. (Value 10.)

Tail.—Long, taper, and nicely curved. (Value 5.)

Coat and Colour.—Coat fine, thick, and close, and colour clear. (Value 5.)

The Greyhound selected to illustrate the breed in this instance is a white and brindled bitch the property of Mr. D. H. Owen, of Belmont Bank, Shrewsbury, and named Lady Shrewsbury. She is a very handsome bitch, a winner at several shows, and also of some coursing stakes, including the cup at Sundorne.

The classes of Greyhounds seen at our shows vary very much as to numbers. There are seldom a dozen specimens at the great London shows, although at the time of the summer show at the Crystal Palace Greyhounds are idle, and could be sent in scores if coursers put any value on prizes won in the show ring. This is not the case, however, and the consequence is that competition is very much limited.

The best classes of Greyhounds are to be met with at provincial shows, in coursing counties, where the local celebrities are shown by their owners; but at many shows one or more good-looking dogs that have been brought out—generally in the North—are first run round a few of the summer shows, and then, getting into the hands of regular exhibitors, snap up most of the prizes throughout the country. Some of these prize dogs have been fair performers, and are eminently handsome specimens, and invariably well bred; I give the measurements of a few, which may be compared with those of some Waterloo Cup Winners, for particulars of which latter I am indebted to the "Coursing Calendar."* The show dogs will have been weighed and measured in a fatter state than the Cup winners, as the weight of the latter represents that which they scaled when trained to run. The Earl of Haddington's Honeywood and Mr. H. G. Miller's Misterton each ran at the weight of 63lb., Princess Dagmar 58lb., Snowflight 47½lb., Wild Mint 45lb., and Coomassie at 42lb., all of these being winners of the Waterloo Cup. In respect to three of them the measurements were as follows:

---

* Published at *The Field* Office, 346, Strand, London.

Measurements.	Princess Dagmar.	Wild Mint.	Honeywood.
Head—	Inches.	Inches.	Inches.
From tip of snout to joining of neck	8½	9	9¾
Girth of head between eyes and ears	13½	13	15¾
„ „ snout...	...	7	9
Distance between eyes	2	1⅞	2
Neck—			
Length from head to shoulders	9½	7¾	9¼
Girth round neck	13¾	12½	14¾
Back—			
From neck to base of tail	24	22	23½
Tail—			
Length	19	17	20
Hips—			
Length of loin from junction of last rib to hip bone	8¼	8	7½
Length of hip bone to socket of thigh joint ...	5½	7½	7½
Fore Leg—			
From base of two middle nails to fetlock joint	...	2½	...
„ fetlock joint to elbow joint	...	10¼	10½
„ elbow joint to shoulder blade	12	11½	13
Thickness of fore leg before the elbow	6½	5¾	7
Hind Leg—			
From hock to stifle joint	12	9¾	10½
Stifle joint, to top of hip bone	12¼	10½	11
Girth of ham part of thigh	16¼	15	17
Thickness of second thigh below stifle	9¼	7	8½
Body—			
Girth round depth of chest	27	26½	30½
„ „ loins	23	18½	23

The following measurements of good show dogs may be taken as a fair average:

Mr. J. L. Bensted's Greyhound *Chimney Sweep*: Age, 5 years; weight, 66lb.; height at shoulder, 26½in.; length from nose to set on of tail, 42¼in.; length of tail, 19in.; girth of chest, 29¾in.; girth of loin, 21in.; girth of head. 15in.; girth of forearm, 6¾in.; length of head from occiput to tip of nose, 10¼in.; girth of muzzle midway between eyes and tip of nose, 8¾in.; measured in working condition. Chimney Sweep won the gold medal in his class at the Paris International Dog Show, 1878.—Mr. J. H. Salter's Greyhound dog *Snapdragon*: Age, 8 years; weight, 72lb.; height at shoulder, 27in.; length from nose to set on of tail, 41in.; length of tail, 19in.; girth of chest, 31¼in.; girth of loin, 22in.; girth of head, 15in.; girth of forearm, 7in.; length of head from occiput to tip of nose, 10½in.; girth of muzzle midway between eyes and tip of nose, 7¾in.—Mr. J. H. Salter's Greyhound bitch *Satanella*: Age, 5 years; weight, 57½lb.; height at shoulder, 24½in.; length from nose to set on of tail, 41½in.; length of tail, 18½in.; girth of chest, 30½in.; girth of loin, 21in.; girth of head, 14½in.; girth of forearm, 6½in.; length of head from occiput to tip of nose, 9in.; girth of muzzle midway between eyes and tip of nose, 8in.

COURSING.

*Antiquity of Coursing—Xenophon's Treatise on Hunting—
Arrian a Master of the Art of Hunting—The Greyhound in
Italy and Greece—Lælaps—Ovid—Cromwell a Courser—
Establishment of the First Coursing Club—Ashdown Park
Meeting — The Malton Club — The Newmarket — The
Amesbury—The Altcar—The Waterloo Cup—Exclusiveness
of the Old Clubs—Influence of Democracy on Coursing—
The National Coursing Club: Constitution of; Officers;
Rules—" Greyhound Stud Book"—Points of the Course—
Definition of Points—Allowances for Accidents—Penalties—
Enclosed Coursing : Principal Grounds for ; Advantages of;
Opinions for and against—Staying Power in Greyhounds—
A Long Run—What Qualities a Coursing Dog should
possess.*

THE hunting of the hare with fleet-footed dogs pursuing their quarry by sight only is of ancient date, but far from being such an old-established sport as the pursuit of the hare and other wild animals by dogs of the sagacious order relying on that faculty of scent which, in many varieties of the dog, is developed to such a marvellous extent as to astonish us, and has led so keen and accurate an observer and cautious reasoner as Haekel to form the opinion that it is not merely our own sense of smell greatly magnified, but one essentially different.

Xenophon, who died at Corinth 395 B.C., at the age of ninety years, wrote a treatise on hunting, the modes of setting snares and nets; the natural history, food, and habits of the hare; how to search for it; and what dogs were clever at scenting, and what faulty; as well as showing how, by stratagem, the fierce animals of the chase—the boar, bear, and lion—might be taken. Hare-hunting by dogs running the hare or stag down by fleetness of foot and keenness of sight was unknown to him, the Greyhound, which was a dog peculiar to the Celts of Gaul and Britain, not having been imported to Italy and Greece till some centuries after Xenophon's time. Nothing in history is much clearer than this fact, and it fully disposes of the specific name of *Grœcus* being used to designate the Greyhound, as indicating for him a Greek origin.

Arrian, who was born at the end of the first or very early in the second century, for he was made a citizen of Rome by the Emperor in 124 A.D., and consequently lived about 500 years later than Xenophon, was a master of the art of hunting in all its then known branches, and well acquainted with the cynegetica of Xenophon, and he took up the subject, adopting the name of his predecessor, and filled up or completed his work by a treatise on coursing with Celtic Greyhounds, which had in the meantime been introduced to Italy and Greece.

In Dansey's translation of this treatise the ignorance of the elder Xenophon of dogs of our Greyhound type, and the thorough knowledge of them possessed by Arrian (or the younger Xenophon), are made evident, this one very short quotation putting the case very clearly. Arrian writes: "And that he (the elder Xenophon) was unacquainted with any breed of dogs resembling the Celtic in point of swiftness is evident from these words: 'Whatever hares are caught by dogs become their prey, contrary to the natural shape of the animal, or accidentally.' Now, if he had been acquainted with the Celtic breed I think he would have made the very same remark on the dogs: 'Whatever hares the dogs do *not* catch at speed, they fail of catching, in contradiction of their shape, or from some accidental circumstance.' For assuredly, when Greyhounds are in good condition, and of high courage, no hare can escape them." This is clear reasoning, by a practical sportsman, and is undeniable. Our Greyhounds are formed for speed; but no dog hunting by scent is cast in the mould to run a hare down by outpacing her.

The introduction of the Greyhound into Italy and Greece had taken place at least a century and a half before the time of Arrian, for the dog is referred to by writers of that period, including Virgil, Ovid, and Gratius. Ovid, in describing the fleetness of the Celtic Hound, poetically identifies the Greyhound with the mythic Lælaps fashioned by Vulcan, and, after changing hands as often as a modern show winner, eventually presented by Procris to Cephalus; but the said Lælaps was described by Pollux as of the Molossian or Mastive type. Ovid's description of a single-handed course, as translated by Dryden, and given on page 20, is almost positive evidence that he was a practical courser; and his description of the dash and swiftness of the Greyhound on being slipped, as rendered by Golding, supplies additional evidence of his having been a partaker in the sport:

> "Scarcely had we let him off from hand,
> But that where Lælaps was we could not understand;
> The print remained of his feet upon the parchèd sand,
> But he was clearly out of sight. Was never dart, I trow,
> Nor pellet from onforced sling, nor shaft from Creetish bow,
> That flew more swift than he did run."

From these ancient days we can trace no important change in the dog or the sport till quite modern times.

Coursing in England was, as already noticed, reduced to a system by the code of laws regulating it drawn up by the Duke of Norfolk in the reign of Elizabeth. From that date the sport grew in favour, and not even the puritanic spirit of the Commonwealth stopped its progress; in fact, Cromwell was himself a courser and an ardent lover of the Greyhound. The next great public act consolidating the sport did not take place till near the end of last century, when the first coursing club was established by Lord Orford, at Swaffham, in 1776. Four years later Lord Craven established the celebrated Ashdown Park Meeting. The following year (1781) saw the Malton Club formed, and gradually such institutions spread over England, the great Newmarket Society being formed in 1805, with numbers of others of more or less importance, up to the celebrated Amesbury in 1822, and the still greater Altcar Society in 1825; and ultimately, in 1836, on the plains of Altcar, the greatest annual coursing event in the world was established, "the blue ribbon of the leash" as it has been called, the Waterloo Cup, now contended for every February by sixty-four Greyhounds, nominated by sixty-four coursers, English, Scotch, and Irish.

The next great event in coursing annals was the establishment of the National Coursing Club in 1858, which began its work by formulating a code of laws for the regulation of coursing meetings and the guidance of judges that at once approved themselves to the great body of coursing men. The laws of the Club are now the laws of the leash, not at home only, but in the Greater Britain beyond the seas, and wherever coursing has taken root, as it has most firmly with our brethren in California, and still more so with our fellow countrymen in Australia.

Long before the establishment of the National Coursing Club causes were at work strengthening the foundations of this national sport and widely extending its influence and popularity, and these, in fact, made the Club a necessity.

The old clubs were, in accordance with the spirit of the times, rigidly exclusive. The votaries of the leash of that period determined that, in coursing at least, "the toe of the peasant should not come so near the heel of the courtier as to gall his kibe," and so we find that no dog was permitted to run at a meeting which was not the property of a member, or of someone taken to the meeting by a member.

But the giant Democracy, having given many uneasy turns in his long sleep, was now rubbing his eyes and stirring in earnest. The days of railways came in with the Waterloo Cup—the first, and still the chief, of open coursing meetings—and the dawn of a really free and independent Press was visible.

These three things—open competition, cheap and rapid travelling, and the Press—brought coursers together and made the doings of all open to all, thus stimulating praiseworthy emulation and giving greater zest to the sport.

In the early days of free coursing, even *Bell's Life* did not give more than the results of stakes and matches; whilst now no sporting paper of repute exists that does not send its specialist to report on the running at all important public meetings, and even the daily papers of London and the chief provincial towns supply their readers with full information respecting the principal coursing events.

The National Coursing Club being the supreme authority on all coursing matters—framing and administering laws, arbitrating on disputes, settling betting questions, and with power to disqualify from participation in the sport at all meetings held under its rules those who infringe or set at nought its rulings—it will be well to glance at its constitution and the basis on which it rests. The Club consists of members elected to that position by the coursing clubs of the United Kingdom which are of more than one year's standing, and which have not less than twenty-four members. The Club thus formed has the power to elect twenty-five additional members from among well-known supporters of public coursing, such members being elected for five years. Ten members from those elected by the several local clubs retire annually, their places being filled by those clubs electing others or re-electing the retiring members. The Club holds two general meetings in the year for the despatch of business, revision or alteration of rules, election of members, &c., one of these in

London on the last Wednesday in June, and the other in Liverpool on the day of entry for the Waterloo Cup.

The Club has instituted a "Greyhound Stud Book," and it is required that every dog that runs at public coursing meetings shall be registered by name, pedigree, and description in that book, which is very ably edited by Mr. D. Brown, Dalry, Ayrshire.

The Earl of Sefton, over whose lands at Altcar the Waterloo Cup is run for, has been for many years president of the Club. The honorary secretary is Mr. R. B. Carruthers, Huntingdon Lodge, Dumfries.

It will be seen from the preceding condensed statement that the National Coursing Club rests on a liberal elective basis, and in this important matter of principle is in direct contrast to the Kennel Club—the ruling body in all matters pertaining to dog shows—which is self-elected.

It is unnecessary here to recapitulate the somewhat lengthy, very complete, and carefully-drawn code of rules of the Club; but what is of the greatest consequence for all to know who would indulge in coursing are the points of the course on which the merits of the competitors are decided by the judge, and these I give *in extenso* here. They are founded on the code of the Duke of Norfolk, and variations on that code made by the earlier clubs, and are the result of the gradual development of coursing under modern conditions, the alterations in them being improvements, in so much that they better meet the wants of to-day. The curious in coursing lore will find the Duke of Norfolk's code of laws in Goodlake, Thacker, and still older writers, and the alterations made in it, traced to that which now rules the sport.

The points of the course are:

(*a.*) *Speed.*—Which shall be estimated as one, two, or three points, according to the degree of superiority shown. [See definition below (*a*).]

(*b.*) *The Go-bye.*—Two points; or, if gained on the outer circle, three points.

(*c.*) *The Turn.*—One point.

(*d.*) *The Wrench.*—Half a point.

(*e.*) *The Kill.*—Two points, or in a descending scale, in proportion to the degree of merit displayed in the kill, which may be of no value.

(*f.*) *The Trip.*—One point.

Definition of points:

(*a.*) In estimating the value of speed to the hare, the judge must take into account the several forms in which it may be displayed, viz.:

1. Where, in the run up, a clear lead is gained by one of the dogs, in which case one, two, or three points may be given, according to the length of the lead, apart from the score for a turn or wrench. In awarding these points the judge shall take into consideration the merits of a lead obtained by a dog which has lost ground at the start, either from being unsighted or from a bad slip, or which has had to run the outer circle.

2. Where one Greyhound leads the other so long as the hare runs straight, but loses the lead from her bending round decidedly in favour of the slower dog of her own accord, in which case the one Greyhound shall score one point for the speed shown, and the other dog score one point for the first turn.

3. Under no circumstances is speed without subsequent work to be allowed to decide a course, except when great superiority is shown by one Greyhound over another in a long lead to covert.

If a dog after gaining the first six points still keeps possession of the hare by superior speed, he shall have double the proscribed allowance for the subsequent points made before his opponent begins to score.

(*b.*) *The Go-bye* is where a Greyhound starts a clear length behind his opponent, and yet passes him in a straight run, and gets a clear length before him.

(*c.*) *The Turn* is where the hare is brought round at not less than a right angle from her previous line.

(*d.*) *The Wrench* is where the hare is bent from her line at less than a right angle; but where she only leaves her line to suit herself, and not from the Greyhound pressing her, nothing is to be allowed.

(*e.*) *The Merits of a Kill* must be estimated according to whether a Greyhound by his own superior dash and skill bears the hare, whether he picks her up through any little accidental circumstances favouring him, or whether she is turned into his mouth, as it were, by the other Greyhound.

(*f.*) *The Trip*, or unsuccessful effort to kill, is where the hare is thrown off her legs, or where a Greyhound flecks her but cannot hold her.

The following allowances shall be made for accidents to a Greyhound during a course; but in every case they shall only be deducted from the other dog's score:

(*a.*) For losing ground at the start, either from being unsighted or from a bad slip, in which case the judge is to decide what amount of allowance is to be made, on the principle that the score of the foremost dog is not to begin until the second has had an opportunity of joining in the course; and the judge may decide the course, or declare the course to be an undecided or no course as he may think fit.

(*b.*) Where a hare bears very decidedly in favour of one of the Greyhounds after the first or subsequent turns, in which case the next point shall not be scored by the dog unduly favoured, or only half his points allowed, according to circumstances. No Greyhound shall receive any allowance for a fall or an accident, with the exception of being ridden over by the owner of the competing Greyhound or his servant, provided for in Rule 30, or when pressing his hare, in which case his opponent shall not count the next point made.

[RULE 30. *Riding over a Greyhound.*—If any subscriber or his servant shall ride over his opponent's Greyhound while running a course, the owner of the dog so ridden over shall, although the course be given against him, be deemed the winner of it, or shall have the option of allowing the dog to remain and run out the stake, and in such a case shall be entitled to half its winnings.]

Penalties are as follow:

(*a.*) Where a Greyhound, from his own defect, refuses to follow the hare at which he is slipped, he shall lose the course.

(*b.*) Where a dog wilfully stands still in a course, or departs from directly pursuing the hare, no points subsequently made by him shall be scored; and if the points made by him up to that time be just equal to those made by his antagonist in the whole course, he shall thereby lose the course; but where one or both dogs stop with the hare in view, through inability to continue the course, it shall be decided according to the number of points gained by each dog during the whole course.

(*c.*) If a dog refuses to fence where the other fences, any points subsequently made by him are not to be scored; but if he does his best to fence, and is foiled by sticking in a meuse, the course shall end there. When the points are equal the superior fencer shall win the course.

The above rules in their present, as in their primitive, form, were intended for the guidance of the sport as carried on in the field, and under natural conditions, so far as the hare was affected; but now they are also applied to a new style of conducting the pastime, which many think unfair to the game, and adverse to the true interests of the sport.

The chief grounds for the carrying on of this system of coursing are at High Gosforth Park, Newcastle-on-Tyne, and Kempton Park, near London. I have not the dimensions of the latter ground, but the former was described in the Prospectus of the Company as about 700yds. long by 100yds. wide. The hares are kept, and artificially fed, in ground surrounded by wire netting 6ft. high; those of them required to be coursed are, the night before the meeting, driven (says the same Prospectus) "into the south-east corner, where four large pens or prisons have been made of wire and brattice cloth, these pens having by artificial means been made to look like young coverts."

This class of coursing receives the support of so many gentlemen of position that one feels bound to think it cannot possibly be so bad in practice as it looks in print; and I have to acknowledge that I have never witnessed such coursing, and I do not care if I never do, for no amount of support, no matter if it is given by the highest in the land, can ever make it better than a distorted imitation and a mere caricature of the ancient and noble sport, thus travestied for purposes of gate money and gambling. It is found profitable at these places to offer very high stakes to be run for, and these secure large entries and the attendance of multitudes of onlookers, the majority of whom probably know nothing of natural coursing, and a large section of whom consist of the "sharps" and "flats" who hang on to the skirts of modern sport, and to whom life without betting would be insipid. The promoters of coursing boxed hares say: "The advantages this description of coursing possesses over that in an open country is very great; there is no tramping after hares, but, on the contrary, trials can be secured as fast as the judge can decide upon them." Let the reader contrast that with the opinion of one of our greatest modern coursers, who was, as well as a sportsman, a gentleman and a scholar: "Coursing, more than any of the other laborious diversions of rural life, while it ministers to our moderate sensual enjoyment, admits also, during the intervals of the actual pursuit, opportunities of conversation with our brethren of the leash, and mental improvement."

Again, consider the hare, and in what contrast does the practice of hunting an animal so timid that the quality has caused her to be named *Lepus timidus*, when turned out of a prison before a crowd, and one in her own chosen form treated as in all coursing I have ever witnessed, and which was recommended by Arrian eighteen centuries ago, and acted on by all sportsmen until this modern innovation debased the sport to suit men who would be coursers in patent leather boots, and the parasites of sport whose only idea of it is that of a medium for enabling them to become possessed of "unearned increment."

Arrian wrote:—"Let the hare creep away from her form as if unperceived, and recover her presence of mind; and then, if she be a racer she will prick up her ears and bound away from her seat with long strides; and the Greyhounds, having capered about as if they were dancing, will stretch out at full speed after her. And at this time is the spectacle worthy indeed of the pains that must necessarily be bestowed on these dogs." To quote an "Amateur of the Leash," writing in the "Sportsman's Cabinet," 1803, respecting a famous match—which did not come off—for a thousand guineas, appointed to be run by Colonel Thornton's Major, brother to the celebrated Snowball, when a boxed hare was coursed in a substituted trial, and killed by Major matched against a small bitch: "Such an exhibition might be bearable to a few sporting amateurs; but could he (Major) have found a tongue when he beheld himself brought to run a hare, turned out of a box, in the month of March, upon Epsom Downs, amid whiskies, buggies, and gingerbread carts, well might he have exclaimed—

'To this complexion am I come at last.'"

That is how our sporting fathers viewed the effeminate Cockney coursing so popular to-day.

One of the greatest qualities the Greyhound could possess in the estimation of past generations of coursing men was staying power. In 1798 a brace of Greyhounds belonging to a gentleman in Carlisle coursed a hare from the Swift, and killed her at Clemmel, a distance of seven miles. In 1800 a brace of Greyhounds in Lincolnshire ran a hare four miles measured in a straight line, and the turns and windings were calculated to make the entire ground covered up to six miles; and many other long courses, from that time down, could be quoted. I contend that to breed Greyhounds for coursing in enclosures of half-a-mile in length is to take the most certain means of destroying one of the most valuable qualities of the breed; and, for that and other reasons referred to, I hold this class of running degrading to the sport which, legitimately followed, is eminently fair to pursuer and pursued, and bracing and re-invigorating to mind and body of the courser.

It is evident that a dog capable of distinguishing himself in such severe contests as coursing in the open frequently entails must possess great speed, stamina, cleverness, and courage. He must also be in a state of very high training, so as to have his muscular powers developed to the utmost, and all the important functional organs—and especially the lungs and heart—in perfect order, unimpeded by fat or the distension of other organs, such as the stomach and liver, as would arise from disease or improper feeding.

It is now for us to consider how such a dog may be obtained. To this end there are, I consider, three essential conditions, with which we may, indeed, fail to produce the very best, but without which we are certain to fail in any endeavour to produce a good one. These are, that the Greyhound shall be well bred, well reared, and well trained, and I think if we take these conditions *seriatim*, and consider each of them in detail, we cannot fail at least to see the way to the making of a good fair Greyhound— one fit for the course and equal competition with his fellows.

BREEDING.

Importance of Securing good Stock—Snowball—Collateral Breeding—Laws of Heredity—Selection of the Brood Bitch—Importance of Pedigree—Trained Brood Bitches—Consanguineous Breeding—Selecting the Dog—Nicking—Season for Breeding—Foster Dams—Œstrum and Service—Predetermining the Sexes—Treatment during Pregnancy—Prevention of Worms in Pups—Transmission and Reproduction of Parasites — Tœnia cœnurus — Vermifuge— Parturition—Pups in the Nest—Weaning.

It may be said of coursing men, with more truth than of most other classes who use dogs, that every one is a breeder. It is at once a greater pleasure and a higher honour to win with dogs a man has bred and reared than with those purchased. Of course there are circumstances which render purchase compulsory if the sport is to be enjoyed at all, but the desire is general to win with dogs that are entirely the courser's own in the sense referred to. It is essentially necessary, therefore, for the young aspirant to coursing fame to pay attention to the principles of breeding as taught by physiology and enforced by experience.

The first step is to secure possession of the right stock,

and, to do so without the direct aid of a friendly adviser, whose experience and judgment can be relied on, such books as the "Coursing Calendar," "Greyhound Stud Book," and other records of coursing must be consulted.

Throughout the history of the sport, since the record of its minutest details began, we find that at succeeding intervals one or more Greyhounds, or a family of them, have exhibited merit so superior to their contemporaries as to have become famous, and with the consequence of a run upon the strain by breeders, resulting in a wide diffusion of that blood, and its commingling with other blood also highly valued by its possessors. Such practice has tended to general improvement, and to the formation of the striking family likeness among Greyhounds which now exists to a degree it never did before, rendering distinction of strains visible only to the eye of the practical connoisseur.

Pedigrees may be said to have come in with the clubs; but even then, and until quite recent times, very strict attention to descent was not paid: especially was the pedigree of the dam neglected, although of equal importance with that of the sire. An excellent instance of this occurs in the pedigree of Snowball, the wonder of his time, and I believe the first Greyhound that was at stud use to the public, which he was from the year 1800 to 1803, if not later, at a fee of £3 3s. His grand-dam was Lord Orford's Czarina, a yellow bitch of unknown pedigree, that won forty-seven matches without being defeated. Snowball was owned by Major Topham, a coursing celebrity, and his dog's fame has been preserved in the lines of Sir Walter Scott:

BREEDING.

> Who knows not Snowball, he whose race renowned
> Is still victorious on each coursing ground?
> Swaffham, Newmarket, and the Roman Camp,
> Have seen them victors o'er each meaner stamp.

The direct and collateral descendants of Snowball (through his famous brother, Major, and sister, Sylvia)—not one of the trio having ever lost a course—were almost equally successful.

To trace back to Snowball—who was, by the way, a pure black dog—was, in the early part of the century, considered an all-sufficient pedigree. In later times, King Cob, Canaradzo, Master McGrath, and others, have occupied the same position, and so will it be with present and future great winners; though there is this difference, that there is an ever-increasing number of known good. dogs to select from.

It does not, however, follow that winners will be equally successful at stud, for although some certainly have been, yet there are many notable failures.

The opinion entertained by many is that the very high training, and the exceedingly severe and sustained efforts the dogs are called upon to make in coursing, renders them less fitting for stud—reduces their prepotency, so that they impart less vigour to their offspring.

Acting on that belief, many Greyhound breeders prefer collateral breeding—that is to say, they would choose as a stud dog the own brother of a great winner in preference to the winner, if the brother was otherwise suitable, and had not been worked at such high pressure. The question of collateral breeding is a most difficult one to decide upon. For my own part, if the dog was not worn out with overwork, and was in a normal state of health, I should, if owning a vigorous bitch, prefer the dog that had actually displayed the qualities I sought to reproduce.

The laws of heredity are not as yet thoroughly understood; but it may be accepted, if not as an axiom, at least as a rule so general in its application as to warrant us in acting upon it, that excellence is more likely to be inherited direct from excellence than from mediocrity, even if of the same blood.

The selection of the brood bitch is of the greatest possible importance. That bitches of inferior or despised pedigrees may have produced good pups, and even in some instances dogs of extraordinary merit, is no argument against the general principle that from the best we get the best. Therefore, choose your bitch as to physical conformation by the description given in a preceding chapter, and let her be of fair size, with a leaning to the big side rather than to the small, and with that build or shape which, applied to mares, is called "roomy"; and be sure she is of a strong, vigorous constitution, with no tendency in her family to develop any special form of disease, for that would suggest an hereditary taint likely to be transmitted to her pups.

The question of pedigree is of the very highest importance; and by pedigree I do not mean a mere string of names. Let the breeder consider the value of the pedigree from the standpoint of merit displayed by the dogs that figure in it, and in doing so note the characteristics that have marked the strain—whether their *forte* has been speed, cleverness, resolution, or endurance—and particularly study the character of the bitch herself in regard to these qualities.

I am presuming that the brood bitch has been trained and used at her proper work, for it is very advisable that every dog of practical use, whether Pointer, Setter, Colley, or Greyhound, should be practised in its proper vocation before being used to the stud; for the secondary instincts created by training and practice are as assuredly hereditary as those we recognise as natural, although, it may be, in a much fainter degree.

The object being to obtain in the pups the qualities above spoken of, the next important step is the selection of a mate. In these days there is abundance to choose from, though it is becoming almost impossible to match a dog and bitch that have not, to a greater or less degree, the same blood in them. It is advisable, however, to avoid too close consanguinity, although, under some circumstances, it may be well to mate the nearest of blood relations, even to brother and sister; but, as may be inferred from the preceding statement, such cases do not now often arise. Still, whenever it is considered advisable to concentrate distinctly-marked and desirable family character, no breeder need hesitate to adopt such close breeding, so long as he does not continue it in-and-in, for that would certainly lead to degeneracy.

The dog and his strain must be considered with care equal to that bestowed on the bitch and her family, alike as to health, physique, and running qualities. As like begets like, within certain limits, and with variations the causes of which are obscure, a dog should be selected possessing — and, if possible, known to inherit — those qualities in which the bitch is deficient. In this way we are more likely to get in the progeny a combination of the qualities required than if, for instance, we mated together those of strains quick for a short course but wanting in bottom. Therefore put the speedy bitch, good only for a spurt, and wanting in staying qualities, to the dog of resolution and endurance, with speed combined; and so in respect to other qualities. This, the only way to correct faults, and to raise mediocrity to the highest level, seems so simple on the face of it, that the desired success in breeding appears easy and certain. As a matter of fact, however, it is far from simple: the laws regulating reproduction and the variations therein are very complex, and disappointments often follow when, according to our best theories, success seems most certain. No rules are absolute, no theories perfect: influences which we cannot see or account for interfere, and theories must be taken as having a general application; the breeder may wisely guide himself by them, but he must neither, on the one hand, be blind to the teachings of his own experience, nor discard a rule, generally true, because he does not in every case attain his object by following it.

In kennel parlance, certain animals when mated are said to "nick"—that is, produce puppies combining the best qualities of both parents — whereas others equally likely to do so, on apparently the soundest theories, fail. Such desirable results may arise fortuitously, but it would be folly to trust to the chapter of accidents when the adoption of intelligible theories give fair promise of success.

The best season for breeding is undoubtedly the spring,

and in my opinion it is much better that pups should not be whelped earlier than March. Spring is the natural breeding season, although under domestication dogs produce young at all seasons of the year—pet and house dogs especially—and even three litters within the year have been recorded. Many owners of dogs like to have litters as early in the year as possible, but I think this is objectionable, for the reason that such have to be protected against the severity of the weather too long; whereas pups whelped in March, April, or May, can have the full benefit of fresh air and sunshine as soon as ever they leave the nest, and nothing tends more than this to encourage a healthy, vigorous growth.

Some bitches are very prolific, producing more pups in one litter than they can rear with justice to themselves and their progeny. To provide against such a contingency, a foster-dam should always be in readiness; one of about the size of the dam is preferable, but the breed is perfectly immaterial. The important point is to be sure she is thoroughly sound and healthy, and not too old, for in the latter case the milk might not be so nourishing and might sooner fail. She should be of a temper to permit of being handled when necessary.

From the first symptoms of œstrum until the bitch has been warded, and through the whole period till the possibility of a *mésalliance* has passed, there is natural anxiety, and extra watchfulness is required lest she should, by any carelessness, be allowed to escape from her confinement and get astray, which she would be almost sure to do, for bitches in season are disposed to stray, and seem to develop at that time a remarkable cunning in eluding their keepers. The best time of heat for the visit is immediately before or after the bleeding, which normally lasts but a few days.

Statements are constantly being made and repeated, to the effect that the sex of the progeny may be regulated at will by the selection of the period at which the bitch visits the dog. Such opinions appear to me to be the offspring of the fruitful imagination of minds plentiful in crops of such weeds because the cultivation of facts has been neglected by them. There are many men who keep dogs and consider themselves breeders on no better grounds than that they have had whelps born in their kennels; and it is this class of men who are always cocksure when others who have studied the subject find themselves in doubt and difficulty. So far as statistics at present collated point, the sexes are produced in nearly equal numbers, with just a small percentage in favour of males, and altogether independent of the time of œstrum at which coition took place.

If conception has taken place there will, after complete disappearance of œstrum, be about two months before the pups are due. The date of service, and that when pups are expected, should be kept in that indispensable requisite for the breeder, the "Kennel Diary,"* which contains a Table showing the interval of sixty-three days—the period of gestation—for every day in the year.

No change from ordinary treatment is required during the earlier period of pregnancy, but as the pupping time approaches care must be taken that the bitch is not

---

* Published by L. Upcott Gill, 170, Strand, London.

allowed, by fencing or any violent exercise, to damage herself and her precious burden. It is safer during the last three weeks to give her exercise in the leash, except in places where she is not likely to be excited to any sudden violent exertion. It is important that the bowels should be acting freely at the time she gives birth, and it is better to secure that by varying the diet than by medicine of any sort; but if the latter must be resorted to, castor oil is the safest.

I have now to suggest a mode of treatment for the prevention of future trouble and loss, which I believe I was the first to advise. It rests entirely on an experience which has been considerable; but by so saying I do not disguise that the treatment is empirical, although I have strong hope that eventually science will show the conjecture on which it rests to have been correct.

Every breeder of Greyhounds knows to his cost that puppies almost as soon as born, if not at and before birth, are infested with worms, which destroy or render useless hundreds; the intestines of puppies only a few days old have been found blocked with them.

Where do they come from? was the question I asked myself; and I asked others—professional men—and received most unsatisfactory answers, mostly involving theories utterly untenable. I consulted such works on animal parasites as were within my reach, learning much, but getting no definite answer to the question that puzzled me. I did, however, from these sources learn that the methods of transmission and reproduction of parasites are as varied and often as marvellous as anything in Nature.

Take, as an example, the well-known case of the *Tænia cœnurus*, one of the tapeworms that infest the dog. Sections of this worm, full of ripe ova, are passed by the dog with his excrement, and scattered on the grass; some of these are swallowed by the sheep; hatched in the stomachs of these animals they find their way to the brain, and there form those bladder worms which cause the well-known disease, "gid." Located there, the *cœnurus* increases by budding, and the luckless dog that afterwards eats the head of the sheep that dies of, or when suffering from, "gid" or "sturdy," very soon has a plentiful crop of tapeworms in his intestines—for thus is the circle of this parasite's life completed. Now, the worms that destroy very young puppies are nematodes, or round worms, of the life history and itinerary of which I am ignorant. But it has occurred to me that as puppies appear to be born with these worms, or their ova, in them, there could be no other source of the parasites than the food supplied to the pups whilst in the uterus.

I have thought it needful to give this somewhat lengthy exposition of my reason for recommending a course which I have no better authority for than my own. I have personally found it answer; but an individual experience is too limited to be taken to settle a question, and of the hundreds of breeders to whom I have recommended the practice through *The Bazaar* newspaper, with which I am connected, very few have taken the trouble to state the result, although such a course would have been very useful to others.

To prevent worms in the future offspring, expel them from the bitch, for it is clear that if there are no

BREEDING.

mature worms there can be no ova to be carried with the nutrient fluids into the systems of the unborn whelps.

A few days after all symptoms of heat have disappeared give a vermifuge, which repeat twice at intervals of about ten days; and if there is any suspicion that the bitch has become re-infested by these parasites, or that the medicine has failed to remove them all, another dose may be given with perfect safety between the fifth and sixth weeks. I would recommend, as the safest and best vermifuges, 1dr. of freshly-grated areca nut, or one of Spratts Patent Worm Powders.

The place set apart for the bitch to whelp in should be retired, so that no other dogs may get near to disturb her, for she would resent the intrusion, with, probably, bad results to herself and pups. The bed should be roomy, very slightly raised from the ground, and have a boarded floor, a low ledge, and hay for litter.

Meddlesome interference is to be deprecated. The Greyhound bitch very rarely requires assistance ; it is the pampered, over-fed show dog that requires the aid of the veterinary accoucheur.

Let the bitch have abundance of food—meat broth and well-boiled oatmeal and milk—and after the first two days take her out for exercise if she does not take it voluntarily.

When the bitch is strong and has sufficient milk, she will let the pups suck her for six or seven weeks; and it is much better to let Nature have its way than to adopt the crotchets of those who are anxious to cram puppies with food before their stomachs are fit to digest it. This practice defeats its object. But to let the puppies have milk or broth to lap at will, without forcing them, gets them on whilst it relieves the dam. Some bitches practise the habit—inherited from wild progenitors—of disgorging half-digested food for their puppies.

As soon as the puppies can toddle about they should have space afforded in which to do so, and that in a sheltered place with choice of sunshine and shade.

Rearing Puppies.

Home Rearing—" Trencher Feeding"—Kennels—Foods and Feeding — Sloppy Foods a Mistake — Exercise—Puppy Troubles: Teething, Worms, External Parasites, Distemper —Remedies—Saplings—Training of Saplings.

We have our choice between home rearing and sending pups out "to walk." The latter plan is very general in the case of Foxhounds, and pups so brought up are termed "trencher fed."

The decision must depend much on the room at command; but in favour of isolated rearing there is the strong argument, so well established by experience, that very large numbers of the young of animals cannot generally be reared together with safety to health. On the other hand, it is not easy to get young dogs placed in the hands of persons at once careful and judicious. Risks, however, must be run in either case.

Where the number of puppies to be raised is not very great, or where puppy kennels of a temporary character can be distributed at distances affording personal supervision over all by the man responsible, I am of opinion that home rearing has the weight of solid reasons in its favour over trusting to the chapter of accidents puppies are open to when in the hands of those who may be well disposed, but have no real knowledge of dogs, and fail to detect when they are going amiss before disease has made serious inroads.

If at walk the puppy has unlimited liberty—which in itself is a good thing, but is not without attendant disadvantages—injuries are more likely to be received, objectionable things are likely to be picked up and eaten, and, unless care is taken, bad habits contracted that are not always easy to eradicate.

If the puppies are, for the most part, kept in kennel, the yard—or, rather, it should be called the playground—ought to be of considerable size, and some contrivance representing miniature hurdles placed across their running ground, so as to compel them, in their romps and chases, to scramble over, and, as they get bigger, to jump, so that the various muscles may be brought into play, which they are not when exercise is confined to the flat.

The kennels ought to be built with a south aspect, and the dormitories should be warm but well ventilated. Absolute cleanliness is an essential to health and growth. It is common to have pointed out, that animals that have been reared amidst dirt, and regardless of all hygienic laws, have grown up strong; but the existence of those who have come well through such an ordeal does not disprove the general law, and those who point to such cases should in fairness also consider the large proportion that have succumbed to unsanitary influences, or grown up weak and puny.

In respect to the diet of growing puppies, I consider it a mistake to feed so much as is usually done on soft, sloppy, almost liquid, food. This when long continued unnecessarily distends the stomach and bowels, and so the evil is to some extent permanent. If we take a line from Nature in this respect we shall not go far amiss. Now the natural habit of the bitch is, before she has weaned her pups, and for some time after, to supply them from her own stomach with half-digested food, which she disgorges for them. This, I think, should teach us the degree of consistence of the food best suited to the pup at that early age. Afterwards the natural food would necessarily become of a firmer character, as the young puppies began to feed on the animals killed for them by the old ones. It is also plain to all, that at the earliest age possible puppies delight to use their teeth by tearing and gnawing bones, &c., and this answers the double purpose of facilitating the irruption of the permanent teeth, and increasing the saliva which aids in digestion; therefore bones of suitable size should be given. The pups should also be accustomed to pieces of Spratts Fibrine Biscuits, dry, as these answer the same purposes referred to as bones do, and are, moreover, highly nutritious.

The first thing in the morning the puppies should have a short run out, to give them an opportunity of relieving themselves, and then at once be fed.

The frequency of the meals must be regulated by the age. When first they leave the dam, four or five times a day is not too often, and whenever they appear hungry between a few bits of dry biscuit may be given. Made food should not be left about in the dishes or troughs from meal to meal, as it either gets fouled, or turns sour, and sets up diarrhœa.

In addition to the biscuits, which should be given both dry and soaked with or boiled in broth, oatmeal, well boiled with milk or broth and a little fat, is good. Indian corn meal and barley meal may be mixed with the oatmeal, but sparingly; and the occasional addition to the food of boiled turnips, carrots, cabbage, or other green vegetables, is beneficial. Flesh is a necessity to the dog as a carnivorous animal.

Where the pups have to be confined to the kennel run for the most part, they must be regularly taken out to the fields and there encouraged to race about; but until they get to seven or eight months old they should not be taken any great distance, and never such as would overtire them.

Young Greyhounds are, perhaps, the most ungainly of dogs, and often the legs look thick and clumsy, the joints being large and prominent. That is no bad sign; but as these bones are yet soft, and scarcely out of the state of gristle, exercise must be given with judgment, according to individual development.

It is the object of breeders of animals of whatever kind to secure the growth of the young without check. This they cannot, however, always succeed in. There are various troubles incident to young life that throw puppies back, and foremost among these is teething, during the period of the irruption of the permanent canines, or fangs, and the molars, or double teeth—that is, from about four to six months old. During this process of Nature the whelp often suffers from feverishness, loss of appetite, and purging. A small dose—about a dessert-spoonful—of castor oil and syrup of buckthorn every other morning for a week, and every other night 3grs. of Dr. James' Fever Powder, will then prove of great benefit.

Worms are the greatest trouble met with in rearing. Puppies suffer from a great variety of tape and round worms, and these are the cause of derangement of stomach and bowels, of severe fits, skin disease, loss of coat, and general unthriftiness; 15grs of areca nut, with 3grs. of santonine, for puppies six months old—others in proportion to age—is a good and safe expellent. It should be given in the morning, and a dessert-spoonful of castor oil an hour after—all before feeding; but warm porridge or broth may be given immediately after the oil.

Of external parasites, there are the invisible mange mite, the flea, louse, and tick, the presence of any one of which is a discredit to the kennelman. White precipitate, rubbed well through the coat and on the skin, and brushed out after a couple of hours, destroys lice and ticks. Washing in quassia water (1oz. of chips boiled for ten minutes in 2galls. of water) kills fleas; and for mange, bi-sulphite of lime, or Spratts' mange lotion, or Bishop's lotion, may be relied on. Many skin diseases arising from functional derangement are mistaken for mange. They require internal remedies, and, unlike true mange, are non-contagious.

The distemper is the other great scourge of the kennel, and is specially a disease of puppies. It is contagious, and so puppies may be easily inoculated with the discharge from nose or eyes of others when it is desired to have all the kennel under treatment at once in preference to the disease lingering on for months. Distemper varies so in the forms it assumes, and requires such different treatment according to its several phases, that I cannot properly discuss the subject here, and must refer the reader to my small book, written for amateurs, "The Diseases of Dogs."*

When the Greyhound is old enough to be slipped at a hare—that is, about ten to twelve months old—he is called a sapling. He does not then properly enter upon his career of coursing; that begins the season following, when he runs through the whole of it as a puppy. There are a few sapling stakes instituted at some coursing meetings, their object being to see and compare the style of running of the youngsters from the different kennels. These are merely trial stakes, and, that the dogs may not be distressed, are limited to four entries; so that at the most a dog can have but two courses, as two of those entered fall in the first round, and the two victors run against each other in the deciding course. Trials of saplings, for the most part, however, take place privately, and it is one of the weeding-out processes of the courser's kennel. The object for which the Greyhound is bred and reared with so much care, and at so much expense, must never be lost sight of. He has not merely to kill his hare—that, indeed, may be said to be but a small part of what is expected from him, and often a Lurcher would do it with more certainty—but he must kill it under strictly-defined conditions, already given; or, if he does not kill, he must have contributed more towards the killing of her than his opponent, to be awarded the victory.

---

* "The Diseases of Dogs: their Pathology, Diagnosis, and Treatment" (L. Upcott Gill, 170, Strand, London).

If, therefore, a sapling displays such gross defects as to make it evident to the judgment of his owner that he does not come up to the standard of a good Greyhound, and cannot, on account of these defects, be reasonably expected to develop qualities required in a good coursing dog, then he is not kept, as it would be folly to incur the very heavy charges of training him.

Saplings before they are thus tried should have some training, for otherwise a courageous youngster getting on to a strong hare might get such a pumper as to do him material injury. Months before the dog can be run as a sapling he should be accustomed to be led both singly and in couples, otherwise, if so handled for the first time at a meeting, he would be most troublesome to his owner, a nuisance to others, and, in fact, probably could not be slipped.

TRAINING.

Every Trainer a Law unto Himself—Nostrums of Bygone Days—Ancient Writers on Training the Greyhound—Systematic Training a Modern Institution — Summer Work—Hardening the Soles : Recipe for—Active Training—Condition of Dogs to be Trained — Physic — Emetics—Vermifuges before Training — Weight — Diet — Food by Weight—Work during Training—Foot Work—Slipping—Test Trials—Grooming—Value of Friction of the Skin—Taking to Meetings — Management during Meetings — Management during Coursing : Rubbing Down—Stimulants.

In the training of Greyhounds every man is a law unto himself, and for the most part the differences in methods are prized in inverse ratio to their importance, mere trifles in practice being very often greatly overvalued. The coursers of to-day have inherited much traditional wisdom on the subject of training, but not unmixed with notions which a clearer appreciation of scientific principles shows us to be unwise. No one nowadays would, I presume, think of training a Greyhound on highly-spiced bread soaked in wine, or on beefsteaks fried in brandy; yet these were practices recommended in bygone times. There still clings to the minds of a certain class of trainers opinions or fads as to Greyhound treatment not quite so gross, but equally untenable. Were it not that trainers are generally men of shrewd common sense, who do not neglect the teachings of their own experience, some of the notions they hold by in theory would sadly mar their calculations. Lessons in the preparation of the Greyhound for the course have come to us from Arrian. In our own tongue De Langley wrote on the subject five hundred years since, and Turberville, Gervase Markham, and others, followed centuries later. In the present century we have had the benefit of the experience, preserved in type, of Goodlake, Thacker, "Stonehenge," and others less known to fame.

Although instructions as to feeding and some general rules as to management are met with in the older writers, yet the severely systematic training now practised is quite modern. Writing in 1803, the author of the "Sportsman's Cabinet," himself a veteran sportsman, and writing over that *nom de plume*, whilst giving credit to Mr. Swinfen's system, which was to give the dogs a brisk gallop twice a day, instead of letting them run at a hare as was the common custom, yet goes on to say: "The training gentlemen of Newmarket, who are universally admitted to know full as much as they ought to do, make no scruple of giving an opinion that training a Greyhound may be brought to as great a certainty as training a horse. The long experienced writer is of a different opinion, feeling himself justified, by ten years' observation, in conceiving these speculative gentlemen are too rapid and uncertain in their declarations."

Practical experience, however, has since conclusively shown that the speculative training gentlemen of Newmarket were right, and that the very charming and well-informed sporting writer I have quoted was, in that instance, wrong.

Although the serious business of training is not to be carried on the year round, yet for saplings, and for older dogs also, attention is constantly needed, and the noble Greyhound must not, during his summer holiday, be allowed, like Artemus Ward's dog, to "lay about and look ornery." Attention to health is always needed, and from the close of the coursing season in March till active training operations in preparation for the opening of the following season in September or October, the dogs should have sufficient of active and regular work to keep them from getting soft in muscle and over fleshy, and also to keep them in discipline.

The dog walks on his toes, and these are an elastic cushion, with a hard, almost horny, covering. Some strains of dogs have naturally better feet than others, but all get soft pads if allowed to lie about idle, or only given exercise on soft turf. The process of hardening the feet is very important and takes a long time, for the harder the tissue the slower the growth.

Regular, but comparatively slow, work over turnpike roads and the like, by following a trap or man on horseback, is the best plan of hardening the pads; but care must be taken not to run the dogs fast where the road has been recently macadamised or covered with flints, or serious, if not permanent, injury may be done.

Dogs with feet naturally soft and prone to spread should have them dipped night and morning in a decoction of oak bark and alum, made in this way: 1lb. of dry oak bark, well bruised, boiled slowly in two gallons of water down to one gallon, and 1oz. of alum dissolved in it. The decoction has to be strained from the bark through a cloth, and may be used over and over again.

Such training as I have spoken of may be considered as merely preliminary and general; but when the date of a coursing meeting is fixed, and certain dogs are selected to be entered and must be ready for the contest, the most active and important work begins. The object is to have the dog, at a given date, in that condition in which he can best display his qualities as a Greyhound. The questions for the courser and the trainer are—What is such condition, and how can it be best attained?

"Stonehenge," universally recognised for a generation past as the highest authority on the subject, with equal strength and terseness says: "Training should not attempt to produce an unnatural condition, but rather the highest state of health."

As already observed, that highest state of health, including the utmost possible physical development with perfect functional order which combined makes possible the acme of vital energy displayed in some given direction, must be attained at a certain fixed date. Now here comes in one of the greatest difficulties and sorest annoyances that afflict the coursing man and his trainer. The night before the coursing is fixed to commence the dogs are perfect; they are all "go," full of life, with flesh firm, and muscles standing out like cords, when Jack Frost steps in and puts the ground in irons, and commits man and dogs to idleness. It is most difficult in these circumstances to keep the dogs up to the mark, for exercise is necessarily limited, and has to be given with care. The best substitute for exercise is increased grooming, and friction along the muscles of locomotion in particular.

The state of the dog in regard to fleshiness and general health has to be carefully taken account of before beginning active training. Dogs then in good health will probably be from seven to ten pounds above their running weight; what the latter should be must be left to the judgment of the trainer, for dogs differ in that as they do in the amount of preparation required.

It is a common practice to begin by giving a brisk dose of purging medicine, such as aloes and jalap. If medicine of the kind is needed, I think 6gr. to 8gr. of compound rhubarb pill with $\frac{1}{4}$gr. of podophyllin and 5gr. of extract of dandelion made into a bolus, is the best form of purgative, as it stimulates the liver as well as relieves the stomach and bowels.

Emetics are frequently given to the dog, and, if judiciously chosen, empty the stomach, and seem to give tone to it. Although vomition is quite a natural act in the dog, for he voluntarily eats couch grass that makes him sick, and apparently with that object, yet strong emetics like tartarised antimony are better avoided. The easiest and safest is ipecacuanha wine, $\frac{1}{2}$oz. to 6dr., given with lukewarm water as a drench, and the dose repeated if vomiting does not follow in ten minutes.

I consider it a better practice than either of the preceding to give worm medicine at this time, and, with that, the bowels must always be acted on by a laxative. Let the bolus prescribed above be given, and the third morning after—the dog being kept to soft food in the

TRAINING. 75

interval—give a worm powder, followed in two hours by ½oz. of castor oil.

The weight should be accurately taken before putting the dog through the mill, and a record of it kept from day to day; this will show how he is progressing better than judging by the eye; but that and the handling of his muscles must also be made factors to our judgment of his condition.

The diet, at all times important, is doubly so now. It is a not uncommon mistake to make too sudden a change from a comparatively poor to a rich diet; such change should be gradually made. Another error is to give a too concentrated food, such as eggs and milk, under the impression that it improves the wind. Either of these courses is likely to cause biliousness, and so upset the dog and retard progress.

Again, much must be trusted to the judgment and experience of the trainer, who, constantly seeing his dogs feed, and how they do on the food, can tell which requires most, and what form of food agrees with one better than another. It is the judicious study and treatment of the individual dog, both as to food and work, that distinguishes the successful trainer. The rule of thumb is, however, never to be safely trusted alone. We must take lessons from the analyst as well as from experience, and select, by the former's researches, the foods most suitable, using the latter in guiding our use of them.

Of flesh we have the choice of beef, mutton, or that of the horse; and without it, in one form or another, the Greyhound cannot be well trained. The choice is often determined by circumstances of convenience; either will do well if sound—that is, if the animal died free of disease, and not saturated with poisonous drugs, which not a few horses whose carcases find their way to kennels are. Of the three, lean mutton is to be preferred, as being more readily digested. It should be prepared by slow boiling; the meat, having been first cut up small, should be put into cold water; let the bones stew with it, and when the meat is well cooked, remove them, and stir in to each pound weight of raw flesh ¼lb. of rough oatmeal, and, after that has boiled slowly half an hour, add ½lb. of Spratts Patent Fibrine Cakes, broken small, and let the whole continue to boil ten minutes longer; that, if a judicious quantity of water has been used (which experience will soon teach), will make the entire mess a stiff pudding when cold enough to feed. Some trainers now use almost entirely Spratts Greyhound Biscuits, containing 50 per cent. of meat, but I think a portion of fresh meat with them preferable.

I have no faith in the practice of doling out food by weight. If a dog, despite all the work that can be given him, gets too fleshy on the diet, reduce its quality rather than its quantity; but the cases are rare where increased exercise will not do the business, by keeping fat off, and that is what is wanted.

On the other hand, if a dog is not doing well on the food, and does not eat it freely, not to say greedily, after he seems indisposed to feed more give him, in pieces, ¼lb. to ½lb. of raw lean mutton, with 10grs. of Boudalt's or Morson's *Pepsin Porci* on it, which is sold in 1oz. bottles, with a measure, at 5s.

There is no need to stick absolutely to the diet above

TRAINING. 77

prescribed, but great and sudden changes are dangerous, and the above I know to be most excellent.

Barley meal is very fattening, and Indian corn meal too heating. If the bowels get constipated, add to the diet boiled bullock's liver, or some boiled carrots or cabbage pressed free from water, rather than resort to medicine.

The value of exercise given by a man on horseback, or simply on foot, and by slipping the dogs, is variously estimated by coursing men. I am of opinion that, when possible, use should be made of the horse, as by that means greater variety of pace is secured. From fifteen to twenty miles a day is not too much, but it is preferable to give it in two instalments — early morning and afternoon—so as, from the latter, to return by feeding time. Before the morning exercise a bit of biscuit should be given, for it is quite as great a mistake to give strong exercise to any animal on an entirely empty stomach as on a full one.

The judgment of the trainer is again called into requisition in regulating the pace to suit the requirements of the dogs, varying it to every degree of speed. He may also have to sort the dogs with different constitutions, taking them in relays, or reserving some for foot exercise and slipping only.

Some men prefer, and others by compulsion have to train the dogs on foot. The practice of reducing the flesh by making the dogs wear heavy clothing is obsolete, and the exercise of the mill as a mode of developing muscle has always, so far as I am aware, been confined to fighting dogs; nor do I think it would be at all suitable to the

Greyhound, any more than it would be to attempt to train a racehorse in a gin. A man could not give on foot the amount of work Greyhounds need, and therefore, if he does not train on horseback, an assistant is needed to slip them. The trainer goes on half a mile or so, and the assistant holds the dogs in leash till he whistles, when, being slipped together, they race to him, the one trying to out-distance the other. It is well in thus slipping to arrange that the dogs shall finish up hill, so as to give them what is called "a good pipe opener." To slip from one hill to finish on another rising ground is very suitable, and if over rough ground—that is, uneven, not stony—so much the better. Some coursing countries being intersected with ditches and fences the Greyhounds should be practised at these, for if unaccustomed to them they would lose by their awkwardness in the actual course.

Where a number have to be slipped, the attendant secures them to his person by a belt round the waist.

On the approach of a great meeting the courser is often puzzled to know which of his dogs to run for certain stakes, and it is customary to put them in slips a week or two before the event with a known good performer; and, when there is none good enough in the home kennel, it is seldom difficult to get this done with the Greyhound of a brother courser, for there is not the secrecy and mystery about these private Greyhound trials that there is in the trial of racehorses, and the "tout" is an unknown nuisance in connection with the sport.

The value of friction of the skin with thorough cleanliness of skin and coat are not often appreciated at their true value, which is very great. In the Greyhound, whilst assisting to keep up a healthy state, judicious rubbing along the principal muscles helps to develop them and to give them increased strength and elasticity.

Night and morning each dog should be taken in hand and well rubbed from neck to end of tail, following the lines of muscle, and rubbing with the hair. Hair gloves are too coarse and irritating for fine-skinned dogs, and a pad of rough towel answers better. Following the neck muscles, those of the shoulder and fore-arms, and down to the feet, must have attention; next along the back, not on the top, but along each side of the spine. The loin muscles, and those of the thigh, down to the hock, are of great importance, and demand to be specially well rubbed, the operator bearing pretty heavily and equally over the whole surface. The best time for this is when the dogs return from exercise.

Where dogs have to travel far by rail or other conveyance, they must be provided with ample room, any cramping being injurious. Such travelling is apt to upset them, and therefore it is well when they are to run in important stakes, to take them to the neighbourhood of the meet a few days in advance; and if the country is of a nature strange to them, it is well to exercise them over land in the neighbourhood similar to it.

The dog should be well exercised the day before the coursing commences, but must not be slipped, or have any fast work, and should not be fed later than two or three o'clock. If the water obtainable is hard, or suspected of being impure, the precaution of boiling as much as the dogs will require should be taken, or they may be upset by it, and so their chances be spoiled.

As the man in charge of the dogs knows in advance how they have been drawn and the order of the running, he should be ready with his dog to put into slips the minute he is wanted.

The dog must be kept clothed to the last; and, when the clothes are removed, a hasty rub down, to stimulate the circulation, should be given just before he goes into slips.

Again, the attendant should keep an eye on his dog from the moment he is slipped, and manage as best he can to follow, so as to be as near as possible at the finish, to at once take his dog up, and so prevent his getting on to a second hare, which in his then partially exhausted state might give him a pumper and very greatly interfere with his chance of winning the courses to follow, as he would be heavily handicapped against an opponent that had not run such extra course.

If the dog has killed, or in running flicked his hare, some of the down may be retained in the mouth, and it must be cleared out; indeed, it is advisable in all cases to sponge the mouth out.

The dog must be gently but well rubbed down till after his breathing returns to its normal state, which will be indicated by his getting his tongue in, and the heavy and rapid movement of the flanks subsiding. The clothing should be put on as soon the rubbing is done, and the dog then kept briskly moving. If very much exhausted, a stimulant may be required; but the administration of any stimulant, especially alcoholic ones, requires great care and judgment, and frequently are much overdone. A dessert-spoonful of brandy is as much as ought to be given to a Greyhound, and this must not be the raw, new spirit, so commonly sold, but old, well-matured Cognac. Milk is a better medium to give it in than water, as it softens the action of the spirit on the coats of the stomach, without reducing its stimulating qualities.

No attempt must be made to give anything until the dog has recovered his breath after a course, or there will be a probability of choking him.

In concluding this Monograph, let me venture to express the hope that its publication may lead to a yet greater interest being taken in the breeding and training of the Greyhound, which, as I have before said, when brought to perfection is the most elegant of the canine race, and the highest achievement of man's skill in manipulating the plastic nature of the dog, and forming it to his special requirements.

I trust that this may be so, and that all who peruse these pages may find something to interest, and induce them to seek the further development of this noble animal according to intelligent and common-sense principles.

As a matter of general interest to all concerned with Greyhounds, I have added as an Appendix a list of the Waterloo Cup Winners from 1836 to 1886.

APPENDIX.

WINNERS OF THE WATERLOO CUP FROM THE COMMENCEMENT.

1836. Mr. Lynn ns (Lord Molyneux's) r b *Milanie*, by Milo out of Duchess.
1837. Mr. Jebb ns (Mr. Stanton's) bk b *Fly*, by Tommy Roads out of Fly.
1838. Mr. Ball's be d *Bugle*, by Bachelor out of Nimble.
1839. Mr. Robinson's r b *Empress*, by Tramp out of Nettle.
1840. Mr. Easterby's bk d *Earwig*, by Hailstone out of Pastime.
1841. Mr. King's r d *Bloomsbury*, by Redcap, Dam by Walton—Sister to Preserve.
1842. Mr. Deakin's f w d *Priam*, by Emperor out of Venus.
1843. Mr. G. Pollok's f d *Major*, by Moses out of Melon.
1844. Mr. N. Slater's r w b *Speculation*, by Sandy out of Enchantress.
1845. Mr. Jebb ns (Mr. Temple's) bk b *Titania*, by Driver out of Zoe.
1846. Mr. Barge ns (Mr. Sampson's) bk w d *Harlequin*, by Emperor out of Lady.
1847. Lord Sefton's r d *Senate*, by Sadek out of Sanctity.
1848. Sir St. G. Gore's bk w b *Shade*, by Nonchalance out of Margery.
1849. Sir St. G. Gore's bk d *Magician*, by King Cob out of Magic.
1850. Mr. G. F. Cooke's f w b *Cerito* (late Lucy Long), by Lingo out of Wanton.
1851. Mr. W. Sharpe's f d *Hughie Graham*, by Liddesdale out of Queen of the May.

WATERLOO CUP WINNERS. 83

1852. Mr. G. F. Cooke's f w b *Cerito*, by Lingo out of Wanton.
1853. Mr. G. F. Cooke's f w b *Cerito*, by Lingo out of Wanton.
1854. Lord Sefton's bk d *Sackcloth*, by Senate out of Cinderella.
1855. Mr. T. Brocklebank ns (Mr. Jefferson's) r d *Judge*, by John Bull out of Fudge.
1856. Mr. J. Bake ns (Mr. W. Peacock's) f b *Protest*, by Weapon out of Pearl.
1857. Mr. W. Wilson's w f d *King Lear*, by Wigan out of Repentance.
1858. Mr. S. Cass's f d *Neville*, by Autocrat out of Catherine Hayes.
1859. Mr. J. Jardine's bk b *Clive*, by Judge out of Mœris. ⎫
    Mr. J. Gordon ns (Mr. J. Jardine's) bk d *Selby* by ⎬ Divided.
    Barrator out of Ladylike. ⎭
1860. Mr. J. Blackstock's r b *Maid of the Mill*, by Judge out of Bartolozzi.
1861. Mr. J. Hyslop ns (Mr. I. Campbell's) w d *Canaradzo*, by Beacon out of Scotland Yet.
1862. Mr. J. Callander ns (Mr. Gregson's) bk b *Roaring Meg*, by Beacon out of Polly.
1863. Mr. T. T. C. Lister's w bk b *Chloe*, by Judge out of Clara.
1864. Mr. T. Williams ns (Dr. Richardson's) w bk d *King Death*, by Canaradzo out of Annoyance.
1865. Col. Goodlake ns (Mr. G. Carruthers') r or f b *Meg*, by Terrona out of Fanny Fickle.
1866. Mr. Gorton ns (Mr. Foulkes') bk w d *Brigadier*, by Boreas out of Wee Nell.
1867. Mr. E. W. Stocker ns (Mr. W. J. Legh's) w bd b *Lobelia*, by Sea Foam out of Lilac.
1868. Lord Lurgan's bk w d p *Master M'Grath*, by Dervock out of Lady Sarah.
1869. Lord Lurgan's bk w d *Master M'Grath*, by Dervock out of Lady Sarah.
1870. Mr. J. Spinks' r b p *Sea Cove* (late Covet), by Strange Idea out of Curiosity.
1871. Lord Lurgan's bk w d *Master M'Grath*, by Dervock out of Lady Sarah.
1872. Mr. J. Briggs' f b *Bed of Stone*, by Portland out of Imperatrice.
1873. Mr. R. Jardine's r w b p *Muriel*, by Fusilier out of Portia.
1874. Mr. C. Morgan's r d *Magnano*, by Cauld Kail out of Isoline.

1875. Mr. W. F. Hutchinson's bk w b *Honeymoon*, by Brigadier out of Hebe.
1876. Mr. R. M. Douglas' bk d *Donald*, by Master Burleigh out of Phœnix.
1877. Mr. R. F. Wilkins ns (Mr. R. Gittus') f w b p *Coomassie*, by Celebrated out of Queen.
1878. Mr. H. F. Stocken ns (Mr. T. Lay's) f w b *Coomassie*, by Celebrated out of Queen.
1879. Mr. H. G. Miller's bk w d p *Misterton*, by Contango out of Lina.
1880. Mr. R. B. Carruthers ns (Earl of Haddington's) r w d *Honeywood*, by Cavalier out of Humming Bird.
1881. Mr. H. G. Miller ns (Mr. J. S. Postle's) w bd b *Princess Dagmar*, by Ptarmigan out of Gallant Foe.
1882. Capt. Ellis ns (Mr. T. Hall's) bk b p *Snowflight*, by Bothal Park out of Curiosity.
1883. Mr. G. J. Alexander ns (Mr. W. Osborne's) r b *Wild Mint*, by Haddo out of Orla.
1884. Mr. C. E. Marfleet ns (Mr. J. Mayer's) w bk d *Mineral Water*, by Memento out of Erzeroum.
1885. Mr. E. Dent's bd w b p *Bit of Fashion*, by Paris out of Pretty Nell.
Mr. J. Hinks ns (Mr. C. Hibbert's) bd b p *Miss Glendyne*, by Paris out of Lady Glendyne. } Divided.
1886. Mr. R. B. Carruthers ns (Mr. C. Hibbert's) bd b *Miss Glendyne*, by Paris out of Lady Glendyne.

INDEX.

A.
	PAGE
Accidents during a course, allowances for	45
Active training	73
Allowances for accidents during a course	45
Arrian, 3, 5, 18, 22, 23, 25, 26, 27, 30, 38, 48, 71	
Ashdown Park Meeting, establishment of	40

B.
Berners, Dame Juliana, on "the properties of a good grehounde"	21
Bitch, brood, selection of	54
food for	61
seclusion of, during pregnancy	61
"Booke of St. Alban's"	6, 21, 29
Breeding	51
collateral	53
consanguineous	55
season for	56
Brood bitches, trained	55
bitch, selection of	54

C.
Caius, Dr. Johannes	7, 11
Celtic hound, the	3, 5
Chaucer	15, 16
Cleanliness an essential for kennels	64
Clubs, coursing, dates of formation of	40
exclusiveness of the old	41
Collateral breeding	53
Colour in greyhounds	30
Condition of dog to be trained	74
Conformation, general, grouping according to	1
Consanguineous breeding	55
Cotton Library, manuscript in	4
Courage, crossing for	32
Course, accidents during, allowances for	45
definition of points of the	44
description of a	18
Ovid's description of a	19
penalties incurred during a	46
points of the	44
Courses, long	49
Coursing	37
antiquity of	13, 37
club, establishment of the first	40
Duke of Norfolk's code of laws for	8
enclosed	46
for wages, first mention of	6
in the time of Elizabeth	7
management of dogs during	80
mistakes of the tyro as to	17
old prejudices against	16
Somerville on	16
Cromwell a courser	40
Crossing with the bulldog	32

D.
Dams, foster	57
Dansey	11, 12
Definition of the points of the course	44
Democracy, influence of, on coursing meetings	42

INDEX.

	PAGE
Derivation of the word greyhound	10
Distemper	67
Dog, selecting the	55

E.

Edward the Confessor and his hounds	5
Elfric, Duke of Mercia, and the greyhound	3
Elizabeth, Queen, as a courser	7
Emetics before training	74
Enclosed coursing	46
"Englishe Dogges," Caius'	7
Exercise during training, substitute for	74
for puppies	64, 65, 66

F.

Feeding, errors of, during training	75
trencher	63
Food by weight, absurdity of giving	76
during training	75
for bitch	61
for pups	61, 64
sloppy, mistake of feeding puppies on	64
sour, evils of	65
Foot work during training	77
Forest laws of Henry III.	6
Foster dams	57

G.

Gazehound, the	7
Gelert	6
Gesner	12
Gestation tables	58
Gratius	30
Grewhound	12
Greyhound, derivation of name	10
modern, elegance of	10
modern, fitnesss of, for designed purpose	10
Stud Book, the	43
uses of, in ancient times	14
Greyhounds, restrictions as to keeping	12
Grooming	79
Group, greyhound, a distinct type	2
Greyhound, general conformation of	1, 14
Grouping according to general conformation	1

H.

	PAGE
Hardening the soles during training	72
Henry III., forest laws of	6
Heredity, laws of	54
Holinshed	3
Home rearing	63
Hunting call, royal, Ossian's description of	5

I.

"Idstone"	30
Introduction of greyhound to Great Britain	3
of the greyhound into Italy and Greece	39

J.

John, King, and his kennels	6
Judging greyhounds	17, 20

K.

Kennels, cleanliness essential for	64
for puppies	64
hygienic conditions necessary for	64
Kill of merit, a	19

L.

Lælaps	39
Langley, Edmund de	6
Laws of heredity	54

M.

Malton Club, formation of	40
Management of dogs during coursing	80
of dogs during meetings	79
Markham, Gervase, 15, 23, 25, 26, 27,	29, 30, 71
"Mayster of Game," 6, 22, 23, 26, 27,	30, 71
Meals for puppies, frequency of	65
Measurements of show dogs	35
of Waterloo Cup winners	35
Meetings, management of dogs during	79
taking greyhounds to	79
Modern greyhound, the	9

N.

National Coursing Club, constitution of	42
Coursing Club, establishment of	41

INDEX.

	PAGE
Newmarket Society, institution of	40
"Nicking"	56
Norfolk's, Duke of, code of laws of the leash	8

O.

Œstrum	57
Oppian	30
Ossian's description of a royal hunting call	5
Ovid	12, 19, 39
Ovid's description of a course	20

P.

Parasites, external, on puppies	67
transmission and reproduction of	59
Parturition	61
Pedigree	52
importance of	54
Penalties incurred during a course	46
Points of the course	44
of the course, definition of	44
of the greyhound, summary of	32
of the greyhound, value of	32
Predetermining the sexes	58
Pregnancy, treatment during	58
Prevention of worms in pups	59
Properties:	
back	27
chest	27
colour	30
courage	32
ears	26
eye	25
head	22
jaws	25
loins	29
muzzle	25
neck	26
size	31
tail	30
teeth	25
weight	31
"Properties of a good grehounde," the	22
Puppies, exercise for	64
food for	64
frequency of meals for	65
kennels for	64
proper consistence of food for	65
rearing	63

	PAGE
Puppy troubles:	
distemper	67
parasites, external	67
teething	66
worms	67
Pups, food for	61
in the nest	61
worms in, prevention of	59
Purgatives before training	74

Q.

Qualities, necessary, in a greyhound	15

R.

Rearing, home	63
puppies	63
Recipe for hardening the soles during training	72
Remedies for puppy troubles,	66, 67
Reproduction of parasites	59
Restrictions as to keeping greyhounds	13
Riding over a greyhound, rule as to	45
Royal Library, picture in	4
Rubbing down	80
Run, a long	49

S.

Sapling, definition of the term	68
trial stakes	63
Saplings, training	69
weeding out	69
Scott, Sir Walter	12, 15
Season for breeding	56
Selecting brood bitch	54
dog	55
Service	57
Sexes, predetermining the	58
Shakespeare	24
Show dogs, measurements of	35
Slipping	78
Sloppy food, mistake of feeding puppies on	64
Snowball, celebrity of	52
Somerville	16
Sour food, evils of	65
Speed of greyhounds, Chaucer on the	15
of greyhounds, Sir W. Scott on the	15
"Sports and Pastimes," Strutt's	4
Staying power	49

88 INDEX.

	PAGE
Stimulants	80
Stock, good, importance of securing	51
"Stonehenge"	12, 30, 71, 73
Summary of points of greyhound	32
Swinfen's, Mr., system of training	71
Systematic training a modern institution	71

T.

	PAGE
Tænia cænurus	60
Teething	66
Test trials during training	78
Trained brood bitches	55
Training	71
active	73
ancient writers on	71
condition of dog before	74
emetics before commencing	74
errors in feeding during	75
food during	75
foot work during	77
grooming	79
hardening the soles	72
horse exercise during	77
management during coursing	80
management during meetings	79
Mr. Swinfen's system of	71
old nostrums used in	70
purgatives before commencing	74
record of weight to be kept during	75
rubbing down	80
saplings	69
slipping	78
stimulants	80
"Stonehenge" on	73
substitute for exercise during	74
Training, summer work	72
systematic, a modern institution	71
taking to meetings	79
test trials during	78
the author of the "Sportsman's Cabinet" on	71
value of friction of the skin during	79
vermifuges before	74
weight, record of, to be kept during	75
work during	77
Transmission of parasites	59
Treatment during pregnancy	58
"Trencher feeding"	63
Trial stakes for saplings	68
Turberville,	7, 18, 30, 71

V.

Value of points of greyhound	32
Vermifuge	61
Vermifuges before training	74

W.

Waterloo Cup, establishment of	40
Cup winners, measurements of	35
Cup winners, list of, from the commencement	82
Weaning	61
Weeding out saplings	69
William of Malmesbury	5
Wolfhound, Irish, strain used for	2
Work during training	77
Worms in puppies	67
in puppies, prevention of	59

X.

Xenophon's treatise on hunting	38

Catalogue of Practical Handbooks Published by L. Upcott Gill, 170, Strand, London, W.C.

ANGLER, BOOK OF THE ALL-ROUND. A Comprehensive Treatise on Angling in both Fresh and Salt Water. In Four Divisions, as named below. By JOHN BICKERDYKE. With over 220 Engravings *In cloth, price 5s. 6d., by post 6s.* (A few copies of a LARGE PAPER EDITION, *bound in Roxburghe, price 25s.*)

 Angling for Coarse Fish. Bottom Fishing, according to the Methods in use on the Thames, Trent, Norfolk Broads, and elsewhere. Illustrated. *In paper, price 1s., by post 1s. 2d.*

 Angling for Pike. The most Approved Methods of Fishing for Pike or Jack. Profusely Illustrated. *In paper, price 1s., by post 1s. 2d.; cloth, 2s. (uncut), by post 2s. 3d.*

 Angling for Game Fish. The Various Methods of Fishing for Salmon; Moorland, Chalk-stream, and Thames Trout; Grayling and Char. Well Illustrated. *In paper, price 1s. 6d., by post 1s. 9d.*

 Angling in Salt Water. Sea Fishing with Rod and Line, from the Shore, Piers, Jetties, Rocks, and from Boats; together with Some Account of Hand-Lining. Over 50 Engravings. *In paper, price 1s., by post 1s. 2d.; cloth, 2s. (uncut), by post 2s. 3d.*

AQUARIA, BOOK OF. A Practical Guide to the Construction, Arrangement, and Management of Fresh-water and Marine Aquaria; containing Full Information as to the Plants, Weeds, Fish, Molluscs, Insects, &c., How and Where to Obtain Them, and How to Keep Them in Health. Illustrated. By REV. GREGORY C. BATEMAN, A.K.C., and REGINALD A. R. BENNETT, B.A. *In cloth gilt, price 5s. 6d., by post 5s. 10d.*

AQUARIA, FRESHWATER: Their Construction, Arrangement, Stocking, and Management. Fully Illustrated. By REV. G. C. BATEMAN, A.K.C. *In cloth gilt, price 3s. 6d., by post 3s. 10d.*

AQUARIA, MARINE: Their Construction, Arrangement, and Management. Fully Illustrated. By R. A. R. BENNETT, B.A. *In cloth gilt, price 2s. 6d., by post 2s. 9d.*

AUSTRALIA, SHALL I TRY? A Guide to the Australian Colonies for the Emigrant Settler and Business Man. With two Illustrations. By GEORGE LACON JAMES. *In cloth gilt, price 3s. 6d., by post 3s. 10d.*

AUTOGRAPH COLLECTING: A Practical Manual for Amateurs and Historical Students, containing ample information on the Selection and Arrangement of Autographs, the Detection of Forged Specimens, &c., &c., to which are added numerous Facsimiles for Study and Reference, and an extensive Valuation Table of Autographs worth Collecting. By HENRY T. SCOTT, M.D., L.R.C.P., &c., Rector of Swettenham, Cheshire. *In leatherette gilt, price 7s. 6d., by post 7s. 10d.*

Published by L. UPCOTT GILL,

BEES AND BEE-KEEPING: Scientific and Practical. By F. R. CHESHIRE, F.L.S., F.R.M.S., Lecturer on Apiculture at South Kensington. *In two vols., cloth gilt, price 16s., by post 16s. 4d.*
 Vol. I., Scientific. A complete Treatise on the Anatomy and Physiology of the Hive Bee. *In cloth gilt, price 7s. 6d., by post 7s. 10d.*
 Vol. II., Practical Management of Bees. An Exhaustive Treatise on Advanced Bee Culture. *In cloth gilt, price 8s. 6d., by post 8s. 10d.*

BEE-KEEPING, BOOK OF. A very practical and Complete Manual on the Proper Management of Bees, especially written for Beginners and Amateurs who have but a few Hives. Fully Illustrated. By W. B. WEBSTER, First-class Expert, B.B.K.A. *In paper, price 1s., by post 1s. 2d.; cloth, 1s. 6d., by post 1s. 8d.*

BEGONIA CULTURE, for Amateurs and Professionals. Containing Full Directions for the Successful Cultivation of the Begonia, under Glass and in the Open Air. Illustrated. By B. C. RAVENSCROFT. *In paper, price 1s., by post 1s. 2d.*

BENT IRON WORK: A Practical Manual of Instruction for Amateurs in the Art and Craft of Making and Ornamenting Light Articles in imitation of the beautiful Mediæval and Italian Wrought Iron Work. By F. J. ERSKINE. Illustrated. *In paper, price 1s., by post 1s. 2d.*

BOAT BUILDING AND SAILING, PRACTICAL. Containing Full Instructions for Designing and Building Punts, Skiffs, Canoes, Sailing Boats, &c. Particulars of the most suitable Sailing Boats and Yachts for Amateurs, and Instructions for their Proper Handling. Fully Illustrated with Designs and Working Diagrams. By ADRIAN NEISON, C.E., DIXON KEMP, A.I.N.A., and G. CHRISTOPHER DAVIES. *In one vol., cloth gilt, price 7s. 6d., by post 7s. 10d.*

BOAT BUILDING FOR AMATEURS, PRACTICAL. Containing Full Instructions for Designing and Building Punts, Skiffs, Canoes, Sailing Boats, &c. Fully Illustrated with Working Diagrams. By ADRIAN NEISON, C.E. Second Edition, Revised and Enlarged by DIXON KEMP, Author of "Yacht Designing," "A Manual of Yacht and Boat Sailing," &c. *In cloth gilt, price 2s. 6d., by post 2s. 9d.*

BOAT SAILING FOR AMATEURS. Containing Particulars of the most Suitable Sailing Boats and Yachts for Amateurs, and Instructions for their Proper Handling, &c. Illustrated with numerous Diagrams. By G. CHRISTOPHER DAVIES. Second Edition, Revised and Enlarged, and with several New Plans of Yachts. *In cloth gilt, price 5s., by post 5s. 4d.*

BOOKBINDING FOR AMATEURS: Being Descriptions of the various Tools and Appliances Required, and Minute Instructions for their Effective Use. By W. J. E. CRANE. Illustrated with 156 Engravings. *In cloth gilt, price 2s. 6d., by post 2s. 9d.*

BUNKUM ENTERTAINMENTS: A Collection of Original Laughable Skits on Conjuring, Physiognomy, Juggling, Performing Fleas, Waxworks, Panorama, Phrenology, Phonograph, Second Sight, Lightning Calculators, Ventriloquism, Spiritualism, &c., to which are added Humorous Sketches, Whimsical Recitals, and Drawing-room Comedies. *In cloth, price 2s. 6d., by post 2s. 9d.*

BUTTERFLIES, THE BOOK OF BRITISH: A Practical Manual for Collectors and Naturalists. Splendidly Illustrated throughout with very accurate Engravings of the Caterpillars, Chrysalids, and Butterflies, both upper and under sides, from drawings by the Author or direct from Nature. By W. J. LUCAS, B.A. *Price 3s. 6d., by post 3s. 9d.*

BUTTERFLY AND MOTH COLLECTING: Where to Search, and What to Do. By G. E. SIMMS. Illustrated. *In paper, price 1s., by post 1s. 2d.*

CACTUS CULTURE FOR AMATEURS: Being Descriptions of the various Cactuses grown in this country; with Full and Practical Instructions for their Successful Cultivation. By W. WATSON, Assistant Curator of the Royal Botanic Gardens, Kew. Profusely Illustrated. *In cloth gilt, price 5s., by post 5s. 3d.*

CAGE BIRDS, DISEASES OF: Their Causes, Symptoms, and Treatment. A Handbook for everyone who keeps a Bird. By DR. W. T. GREENE, F.Z.S. *In paper, price 1s., by post 1s. 2d.*

CAGE BIRDS, BRITISH. Containing Full Directions for Successfully Breeding, Rearing, and Managing the various British Birds that can be kept in Confinement. Illustrated with COLOURED PLATES and numerous finely-cut Wood Engravings. By R. L. WALLACE. *In cloth gilt, price 10s. 6d., by post 10s. 10d.*

CANARY BOOK. The Breeding, Rearing, and Management of all Varieties of Canaries and Canary Mules, and all other matters connected with this Fancy. By ROBERT L. WALLACE. Third Edition. *In cloth gilt, price 5s., by post 5s. 4d.; with COLOURED PLATES, 6s. 6d., by post 6s. 10d.;* and as follows:

General Management of Canaries. Cages and Cage-making Breeding, Managing, Mule Breeding, Diseases and their Treatment, Moulting, Pests, &c. Illustrated. *In cloth, price 2s. 6d., by post 2s. 9d.*

Exhibition Canaries. Full Particulars of all the different Varieties, their Points of Excellence, Preparing Birds for Exhibition, Formation and Management of Canary Societies and Exhibitions. Illustrated. *In cloth, price 2s. 6d., by post 2s. 9d.*

CANOE BUILDING FOR AMATEURS: A Practical Manual, with Plans, Working Diagrams, and full Instructions. By COTTERILL SCHOLEFIELD. *Price 2s. 6d., by post 2s. 9d.* [*In the Press.*

CARD TRICKS, BOOK OF, for Drawing-room and Stage Entertainments by Amateurs; with an exposure of Tricks as practised by Card Sharpers and Swindlers. Numerous Illustrations. By PROF. R. KUNARD. *In illustrated wrapper, price 2s. 6d., by post 2s. 9d.*

CATS, DOMESTIC OR FANCY: A Practical Treatise on their Antiquity, Domestication, Varieties, Breeding, Management, Diseases and Remedies, Exhibition and Judging. By JOHN JENNINGS. Illustrated. *In cloth, price 2s. 6d., by post 2s. 9d.*

CHRYSANTHEMUM CULTURE, for Amateurs and Professionals. Containing Full Directions for the Successful Cultivation of the Chrysanthemum for Exhibition and the Market. Illustrated. By B. C. RAVENSCROFT. *In paper, price 1s., by post 1s. 2d.*

COINS, A GUIDE TO ENGLISH PATTERN, in Gold, Silver, Copper, and Pewter, from Edward I. to Victoria, with their Value. By the REV. G. F. CROWTHER, M.A. Illustrated. *In silver cloth, with gilt facsimiles of Coins, price 5s., by post 5s. 3d.*

COINS OF GREAT BRITAIN AND IRELAND, A GUIDE TO THE, in Gold, Silver and Copper, from the Earliest Period to the Present Time, with their Value. By the late Colonel W. STEWART THORBURN. With 27 Plates in Gold, Silver, and Copper, and 8 Plates of Gold and Silver Coins in RAISED FACSIMILE. *In cloth, with silver facsimiles of Coins, price 7s. 6d., by post 7s. 10d.*

COLLIE, THE. Its History, Points, and Breeding. By HUGH DALZIEL. Illustrated with Coloured Frontispiece and Plates. *In paper, price 1s., by post 1s. 2d.; cloth, 2s., by post 2s. 3d.*

COLLIE STUD BOOK. Edited by HUGH DALZIEL. *Price 3s. 6d. each, by post 3s. 9d. each.*

Vol. I., containing Pedigrees of 1308 of the best-known Dogs, traced to their most remote known ancestors; Show Record to Feb., 1890, &c.
Vol. II. Pedigrees of 795 Dogs, Show Record, &c.
Vol. III. Pedigrees of 786 Dogs, Show Record, &c.

COLUMBARIUM, MOORE'S. Reprinted Verbatim from the original Edition of 1735, with a Brief Notice of the Author. By W. B. TEGETMEIER, F.Z.S., Member of the British Ornithologists' Union. *Price 1s., by post 1s. 2d.*

CONJURING, BOOK OF MODERN. A Practical Guide to Drawing-room and Stage Magic for Amateurs. By PROFESSOR R. KUNARD. Illustrated. *In illustrated wrapper, price 2s. 6d., by post 2s. 9d.*

COOKERY FOR AMATEURS; or, French Dishes for English Homes of all Classes. Includes Simple Cookery, Middle-class Cookery, Superior Cookery, Cookery for Invalids, and Breakfast and Luncheon Cookery. By MADAME VALÉRIE. Second Edition. *In paper, price 1s., by post 1s. 2d.*

CUCUMBER CULTURE FOR AMATEURS. Including also Melons, Vegetable Marrows, and Gourds. Illustrated. By W. J. MAY. *In paper, price 1s., by post 1s. 2d.*

CYCLIST'S ROUTE MAP of England and Wales. The Third Edition; thoroughly Revised. Shows clearly all the Main, and most of the Cross, Roads, and the Distances between the Chief Towns, as well as the Mileage from London. In addition to this, Routes of *Thirty of the most Interesting Tours* are printed in red. The map is mounted on linen, and is the fullest, handiest, and best tourist's map in the market. *In cloth, price 1s., by post 1s. 2d.*

DOGS, BREAKING AND TRAINING: Being Concise Directions for the proper education of Dogs, both for the Field and for Companions. Second Edition. By "PATHFINDER." With Chapters by HUGH DALZIEL. Illustrated. *In cloth gilt, price 6s. 6d., by post 6s. 10d.*

DOGS, BRITISH, ANCIENT AND MODERN: Their Varieties, History, and Characteristics. By HUGH DALZIEL, assisted by Eminent Fanciers. SECOND EDITION, Revised and Enlarged. Illustrated with First-class COLOURED PLATES and full-page Engravings of Dogs of the Day. This is the fullest work on the various breeds of dogs kept in England. In three volumes, *demy 8vo, cloth gilt, price 10s. 6d. each, by post 11s. 1d. each.*

Dogs Used in Field Sports. Containing Particulars of the following among other Breeds: Greyhound, Irish Wolfhound, Bloodhound, Foxhound, Harrier, Basset, Dachshund, Pointer, Setters, Spaniels, and Retrievers. SEVEN COLOURED PLATES and 21 full-page Engravings.

Dogs Useful to Man in other Work than Field Sports; House and Toy Dogs. Containing Particulars of the following, among other Breeds: Collie, Bulldog, Mastiff, St. Bernards, Newfoundland, Great Dane, Fox and all other Terriers, King Charles and Blenheim Spaniels, Pug, Pomeranian, Poodle, Italian Greyhound, Toy Dogs, &c., &c. COLOURED PLATES and full-page Engravings.

Practical Kennel Management: A Complete Treatise on all Matters relating to the Proper Management of Dogs, whether kept for the Show Bench, for the Field, or for Companions. Illustrated with Coloured and numerous other Plates. [*In the Press.*

DOGS, DISEASES OF: Their Causes, Symptoms, and Treatment; Modes of Administering Medicines; Treatment in cases of Poisoning, &c. For the use of Amateurs. By HUGH DALZIEL. Third Edition. *In paper, price 1s., by post 1s. 2d.; in cloth gilt, 2s., by post 2s. 3d.*

ENTERTAINMENTS, AMATEUR, FOR CHARITABLE AND OTHER OBJECTS: How to Organize and Work them with Profi and Success. By ROBERT GANTHONY. *In coloured cover, price 1s., by post 1s. 2d.*

FANCY WORK SERIES, ARTISTIC. A Series of Illustrated Manuals on Artistic and Popular Fancy Work of various kinds. Each number is complete in itself, and issued at the uniform *price of 6d., by post 7d.* Now ready—(1) MACRAMÉ LACE (Second Edition); (2) PATCH-WORK; (3) TATTING; (4) CREWEL WORK; (5) APPLIQUÉ; (6) FANCY NETTING.

FERNS, THE BOOK OF CHOICE: for the Garden, Conservatory, and Stove. Describing the best and most striking Ferns and Selaginellas, and giving explicit directions for their Cultivation, the formation of Rockeries, the arrangement of Ferneries, &c. By GEORGE SCHNEIDER. With numerous Coloured Plates and other Illustrations. *In 3 vols., large post 4to. Cloth gilt, price £3 3s., by post £3 6s.*

FERNS, CHOICE BRITISH. Descriptive of the most beautiful Variations from the common forms, and their Culture. By C. T. DRUERY, F.L.S. Very accurate PLATES, and other Illustrations. *In cloth gilt, price 2s. 6d., by post 2s. 9d.*

FERRETS AND FERRETING. Containing Instructions for the Breeding, Management, and Working of Ferrets. Second Edition, Rewritten and greatly Enlarged. Illustrated. *In paper, price 6d., by post 7d.*

FERTILITY OF EGGS CERTIFICATE. These are Forms of Guarantee given by the Sellers to the Buyers of Eggs for Hatching, undertaking to refund value of any unfertile eggs, or to replace them with good ones. Very valuable to sellers of eggs, as they induce purchases. *In books, with counterfoils, price 6d., by post 7d.*

FIREWORK-MAKING FOR AMATEURS. A complete, accurate, and easily-understood work on Making Simple and High-class Fireworks. By Dr. W. H. BROWNE, M.A. *In paper, price 2s. 6d., by post 2s. 9d.*

FOREIGN BIRDS, FAVOURITE, for Cages and Aviaries. How to Keep them in Health. Fully Illustrated. By W. T. GREENE, M.A., M.D., F.Z.S., &c. *In cloth, price 2s. 6d., by post 2s. 9d.*

FOX TERRIER, THE. Its History, Points, Breeding, Rearing, Preparing, for Exhibition, and Coursing. By HUGH DALZIEL. Illustrated with Coloured Frontispiece and Plates. *In paper, price 1s., by post 1s. 2d.; cloth, 2s., by post 2s. 3d.*

FOX TERRIER STUD BOOK. Edited by HUGH DALZIEL. *Price 3s. 6d. each., by post 3s. 9d. each.*

Vol. I., containing Pedigrees of over 1400 of the best-known Dogs, traced to their most remote known ancestors.

Vol. II. Pedigrees of 1544 Dogs, Show Record, &c.
Vol. III. Pedigrees of 1214 Dogs, Show Record, &c.
Vol. IV. Pedigrees of 1168 Dogs, Show Record, &c.
Vol. V. Pedigrees of 1662 Dogs, Show Record, &c.

FRETWORK AND MARQUETRY. A Practical Manual of Instructions in the Art of Fret-cutting and Marquetry Work. By D. DENNING. *In cloth, price 2s. 6d., by post 2s. 10d.*

FRIESLAND MERES, A CRUISE ON THE. By ERNEST R. SUFFLING. Illustrated. *In paper, price 1s., by post 1s. 2d.*

GAME AND GAME SHOOTING, NOTES ON. Grouse, Partridges, Pheasants, Hares, Rabbits, Quails, Woodcocks, Snipe, and Rooks. By J. J. MANLEY. Illustrated. *In cloth gilt, price 7s. 6d., by post 7s. 10d.*

GAME PRESERVING, PRACTICAL. Containing the fullest Directions for Rearing and Preserving both Winged and Ground Game, and Destroying Vermin; with other Information of Value to the Game Preserver. Illustrated. By WILLIAM CARNEGIE. *In cloth gilt, demy 8vo, price 21s., by post 21s. 9d.*

GARDENING, DICTIONARY OF. A Practical Encyclopædia of Horticulture, for Amateurs and Professionals. Illustrated with 2440 Engravings. Edited by G. NICHOLSON, Curator of the Royal Botanic Gardens, Kew; assisted by Prof. Trail, M.D., Rev. P. W. Myles, B.A., F.L.S., W. Watson, J. Garrett, and other Specialists. *In 4 vols., large post 4to. In cloth gilt, price £3, by post £3 3s.*

GOAT, BOOK OF THE. Containing Full Particulars of the various Breeds of Goats, and their Profitable Management. With many Plates. By H. STEPHEN HOLMES PEGLER. Third Edition, with Engravings and Coloured Frontispiece. *In cloth gilt, price 4s. 6d., by post 4s. 10d.*

GOAT-KEEPING FOR AMATEURS: Being the Practical Management of Goats for Milking Purposes. Abridged from "The Book of the Goat." Illustrated. *In paper, price 1s., by post 1s. 2d.*

GRAPE GROWING FOR AMATEURS. A Thoroughly Practical Book on Successful Vine Culture. By E. MOLYNEUX. Illustrated. *In paper, price 1s., by post 1s. 2d.*

GREENHOUSE MANAGEMENT FOR AMATEURS. The Best Greenhouses and Frames, and How to Build and Heat them, Illustrated Descriptions of the most suitable Plants, with general and Special Cultural Directions, and all necessary information for the Guidance of the Amateur. Second Edition, Revised and Enlarged. Magnificently Illustrated. By W. J. MAY. *In cloth gilt, price 5s., by post 5s. 4d.*

GREYHOUND, THE: Its History, Points, Breeding, Rearing, Training, and Running. By HUGH DALZIEL. With Coloured Frontispiece. *In cloth gilt, demy 8vo, price 2s. 6d., by post 2s. 9d.*

GUINEA PIG, THE, for Food, Fur, and Fancy. Illustrated with Coloured Frontispiece and Engravings. An exhaustive book on the Varieties of the Guinea Pig, and its Management. By C. CUMBERLAND, F.Z.S. *In cloth gilt, price 2s. 6d., by post 2s. 9d.*

HAND CAMERA MANUAL, THE. A Practical Handbook on all Matters connected with the Use of the Hand Camera in Photography. Illustrated. By W. D. WELFORD. Second Edition. *Price 1s., by post 1s. 2d.*

HANDWRITING, CHARACTER INDICATED BY. With Illustrations in Support of the Theories advanced taken from Autograph Letters of Statesmen, Lawyers, Soldiers, Ecclesiastics, Authors, Poets, Musicians, Actors, and other persons. Second Edition. By R. BAUGHAN. *In cloth gilt, price 2s. 6d., by post 2s. 9d.*

HARDY PERENNIALS and Old-fashioned Garden Flowers. Descriptions, alphabetically arranged, of the most desirable Plants for Borders, Rockeries, and Shrubberies, including Foliage as well as Flowering Plants. Profusely Illustrated. By J. WOOD. *In cloth, price 5s., by post 5s. 4d.*

HAWK MOTHS, BOOK OF BRITISH. A Popular and Practical Manual for all Lepidopterists. Copiously illustrated in both colours and black and white from drawings from Nature by the Author. By W. J. LUCAS, B.A. *[In the Press.*

HOME MEDICINE AND SURGERY: A Dictionary of Diseases and Accidents, and their proper Home Treatment. For Family Use. By W. J. MACKENZIE, M.D. Illustrated. *In cloth, price 2s. 6d., by post 2s. 9d.*

HORSE-KEEPER, THE PRACTICAL. By GEORGE FLEMING, C.B., LL.D., F.R.C.V.S., late Principal Veterinary Surgeon to the British Army, and Ex-President of the Royal College of Veterinary Surgeons. *In cloth, price 3s. 6d., by post 3s. 10d.*

HORSE-KEEPING FOR AMATEURS. A Practical Manual on the Management of Horses, for the guidance of those who keep one or two for their personal use. By FOX RUSSELL. *In paper, price 1s., by post 1s. 2d.; cloth, 2s., by post 2s. 3d.*

HORSES, DISEASES OF: Their Causes, Symptoms, and Treatment. For the use of Amateurs. By HUGH DALZIEL. *In paper, price 1s., by post 1s. 2d.; cloth 2s., by post 2s. 3d.*

INLAND WATERING PLACES. A Description of the Spas of Great Britain and Ireland, their Mineral Waters, and their Medicinal Value, and the attractions which they offer to Invalids and other Visitors. Profusely illustrated. A Companion Volume to "Seaside Watering Places." *In cloth, price 2s. 6d., by post 2s. 10d.*

JOURNALISM, PRACTICAL: How to Enter Thereon and Succeed. A book for all who think of "writing for the Press." By JOHN DAWSON. *In cloth gilt, price 2s. 6d., by post 2s. 9d.*

LAYING HENS, HOW TO KEEP and to Rear Chickens in Large or Small Numbers, in Absolute Confinement, with Perfect Success. By MAJOR G. F. MORANT. *In paper, price 6d., by post 7d.*

LEGAL PROFESSION, A GUIDE TO THE. How to Enter either Branch of the Legal Profession, with a Course of Study for each of the Examinations, and selected Papers of Questions. By J. H. SLATER, Barrister-at-Law. *In cloth, price 7s. 6d., by post 7s. 10d.*

LIBRARY MANUAL, THE. A Guide to the Formation of a Library, and the Values of Rare and Standard Books. By J. H. SLATER, Barrister-at-Law. Third Edition. Revised and Greatly Enlarged. *In cloth gilt, price 7s. 6d., by post 7s. 10d.*

MICE, FANCY: Their Varieties, Management, and Breeding. Re-issue, with Criticisms and Notes by DR. CARTER BLAKE. Illustrated. *In paper, price 6d., by post 7d.*

MILLINERY, HANDBOOK OF. A Practical Manual of Instruction. Illustrated. By MME. ROSÉE, Court Milliner, Principal of the School of Millinery. *In paper, price 1s., by post 1s. 2d.*

MODEL YACHTS AND BOATS: Their Designing, Making, and Sailing. Illustrated with 118 Designs and Working Diagrams. A splendid book for boys and others interested in making and rigging toy boats for sailing. It is the best book on the subject now published. By J. DU V. GROSVENOR. *In leatherette, price 5s., by post 5s. 3d.*

MONKEYS, PET, and How to Manage Them. Illustrated. By ARTHUR PATTERSON. *In cloth gilt, price 2s. 6d., by post 2s. 9d.*

RABBIT, BOOK OF THE. A Complete Work on Breeding and Rearing all Varieties of Fancy Rabbits, giving their History, Variations, Uses, Points, Selection, Mating, Management, &c., &c. SECOND EDITION. Edited by KEMPSTER W. KNIGHT. Illustrated with Coloured and other Plates. *In cloth gilt, price* 10s. 6d., *by post* 11s.

RABBITS, DISEASES OF: Their Causes, Symptoms, and Cure. With a Chapter on THE DISEASES OF CAVIES. Reprinted from "The Book of the Rabbit" and "The Guinea Pig for Food, Fur, and Fancy." *In paper, price* 1s., *by post* 1s. 2d.

RABBIT-FARMING, PROFITABLE. A Practical Manual, showing how Hutch Rabbit-farming in the Open can be made to Pay Well. By MAJOR G. F. MORANT. *In paper, price* 1s., *by post* 1s. 2d.

RABBITS FOR PRIZES AND PROFIT. The Proper Management of Fancy Rabbits in Health and Disease, for Pets or the Market, and Descriptions of every known Variety, with Instructions for Breeding Good Specimens. Illustrated. By CHARLES RAYSON. *In cloth gilt, price* 2s. 6d., *by post* 2s. 9d. Also in Sections, as follows:—

General Management of Rabbits. Including Hutches, Breeding, Feeding, Diseases and their Treatment, Rabbit Courts, &c. Fully Illustrated. *In paper, price* 1s., *by post* 1s. 2d.

Exhibition Rabbits. Being descriptions of all Varieties of Fancy Rabbits, their Points of Excellence, and how to obtain them. Illustrated. *In paper, price* 1s., *by post* 1s. 2d.

REPOUSSÉ WORK FOR AMATEURS: Being the Art of Ornamenting Thin Metal with Raised Figures. By L. L. HASLOPE. Illustrated. *In cloth gilt, price* 2s. 6d., *by post* 2s. 9d.

ROSES FOR AMATEURS. A Practical Guide to the Selection and Cultivation of the best Roses. Illustrated. By the REV. J. HONYWOOD D'OMBRAIN, Hon. Sec. of the Nat. Rose Soc. *In paper, price* 1s., *by post* 1s. 2d.

SAILING GUIDE TO THE SOLENT AND POOLE HARBOUR, with Practical Hints as to Living and Cooking on, and Working a Small Yacht. By LIEUT.-COLONEL T. G. CUTHELL. Illustrated with Coloured Charts. *In cloth, price* 2s. 6d., *by post* 2s. 9d.

SAILING TOURS. The Yachtman's Guide to the Cruising Waters of the English and Adjacent Coasts. By FRANK COWPER, B.A.

Vol. I., the Coasts of Essex and Suffolk, containing Descriptions of every Creek from the Thames to Aldborough. Numerous Charts and Illustrations. *In cloth, price* 5s., *by post* 5s. 3d.

Vol. II. The South Coast, from the Thames to the Scilly Islands, with twenty-five Charts printed in Colours. *In cloth, price* 7s. 6d., *by post* 7s. 10d.

Vol. III. The Coast of Brittany: Descriptions of every Creek, Harbour, and Roadstead from L'Abervrach to St. Nazaire, and an Account of the Loire. With twelve Charts, printed in Colours. *In crown 8vo, cloth gilt, price* 7s. 6d., *by post* 7s. 10d.

Vol. IV. The West Coast. Including the East Coast of Ireland. [*In the Press.*

ST. BERNARD, THE. Its History, Points, Breeding, and Rearing. By HUGH DALZIEL. Illustrated with Coloured Frontispiece and Plates. *In cloth, price* 2s. 6d., *by post* 2s. 9d.

ST. BERNARD STUD BOOK. Edited by HUGH DALZIEL. Price 3s. 6d. each., *by post* 3s. 9d. each.

Vol. I. Pedigrees of 1278 of the best known Dogs, traced to their most remote known ancestors, Show Record, &c.

Vol. II. Pedigrees of 564 Dogs, Show Record, &c.

SEA-FISHING FOR AMATEURS. Practical Instructions to Visitors at Seaside Places for Catching Sea-Fish from Pier-heads, Shore, or Boats, principally by means of Hand Lines, with a very useful List of Fishing Stations, the Fish to be caught there, and the Best Seasons. By FRANK HUDSON. Illustrated. *In paper, price 1s., by post 1s. 2d.*

SEA-FISHING ON THE ENGLISH COAST. The Art of Making and Using Sea-Tackle, with a full account of the methods in vogue during each month of the year, and a Detailed Guide for Sea-Fishermen to all the most Popular Watering Places on the English Coast. By F. G. AFLALO. Illustrated. *In cloth, price 2s. 6d., by post 2s. 9d.*

SEASIDE WATERING PLACES. A Description of the Holiday Resorts on the Coasts of England and Wales, the Channel Islands, and the Isle of Man, giving full particulars of them and their attractions, and all information likely to assist persons in selecting places in which to spend their Holidays according to their individual tastes. Illustrated. Seventh Edition. *In cloth, price 2s. 6d., by post 2s. 10d.*

SHADOW ENTERTAINMENTS, and How to Work Them. By A. PATTERSON. *[In the Press.*

SHAVE, AN EASY: The Mysteries, Secrets, and Whole Art of, laid bare for 1s., *by post 1s. 2d.* Edited by JOSEPH MORTON.

SHEET METAL, WORKING IN: Being Practical Instructions for Making and Mending Small Articles in Tin, Copper, Iron, Zinc, and Brass. Illustrated. Third Edition. By the Rev. J. LUKIN, B.A. *In paper, price 1s., by post 1s. 1d.*

SHORTHAND, ON GURNEY'S SYSTEM (IMPROVED), LESSONS IN: Being Instructions in the Art of Shorthand Writing as used in the Service of the two Houses of Parliament. By R. E. MILLER. *In paper, price 1s., by post 1s. 2d.*

SHORTHAND, EXERCISES IN, for Daily Half Hours, on a Newly-devised and Simple Method, free from the Labour of Learning. Illustrated. Being Part II. of "Lessons in Shorthand on Gurney's System (Improved)." By R. E. MILLER. *In paper, price 9d., by post 10d.*

SHORTHAND SYSTEMS; WHICH IS THE BEST? Being a Discussion, by various Experts, on the Merits and Demerits of all the principal Systems, with Illustrative Examples. Edited by THOMAS ANDERSON. *In paper, price 1s., by post 1s. 2d.*

SKATING CARDS: An Easy Method of Learning Figure Skating, as the Cards *can be used on the Ice. In cloth case, 2s. 6d., by post 2s. 9d.;* leather, 3s. 6d., by post 3s. 9d. A cheap form is issued printed on paper and made up as a small book, 1s., *by post 1s. 1d.*

SLEIGHT OF HAND. A Practical Manual of Legerdemain for Amateurs and Others. New Edition, Revised and Enlarged. Profusely Illustrated. By E. SACHS. *In cloth gilt, price 6s. 6d., by post 6s. 10d.*

TAXIDERMY, PRACTICAL. A Manual of Instruction to the Amateur in Collecting, Preserving, and Setting-up Natural History Specimens of all kinds. With Examples and Working Diagrams. By MONTAGU BROWNE, F.Z.S., Curator of Leicester Museum. Second Edition. *In cloth gilt, price 7s. 6d., by post 7s. 10d.*

THAMES GUIDE BOOK. From Lechlade to Richmond. For Boating Men, Anglers, Picnic Parties, and all Pleasure-seekers on the River. Arranged on an entirely new plan. Second Edition, profusely illustrated. *In paper, price* 1s., *by post* 1s. 3d.; *cloth*, 1s. 6d., *by post* 1s. 9d.

TOMATO AND FRUIT GROWING as an Industry for Women. Lectures given at the Forestry Exhibition, Earl's Court, during July and August, 1893. By GRACE HARRIMAN, Practical Fruit Grower and County Council Lecturer. *In paper, price* 1s., *by post* 1s. 1d.

TOMATO CULTURE FOR AMATEURS. A Practical and very Complete Manual on the Subject. By B. C. RAVENSCROFT. Illustrated. *In paper, price* 1s., *by post* 1s. 3d.

TRAPPING, PRACTICAL: Being some Papers on Traps and Trapping for Vermin, with a Chapter on General Bird Trapping and Snaring. By W. CARNEGIE. *In paper, price* 1s., *by post* 1s. 2d.

TURNING FOR AMATEURS: Being Descriptions of the Lathe and its Attachments and Tools, with Minute Instructions for their Effective Use on Wood, Metal, Ivory, and other Materials. Second Edition, Revised and Enlarged. By JAMES LUKIN, B.A. Illustrated with 144 Engravings. *In cloth gilt, price* 2s. 6d., *by post* 2s. 9d.

TURNING LATHES. A Manual for Technical Schools and Apprentices. A guide to Turning, Screw-cutting, Metal-spinning, &c. Edited by JAMES LUKIN, B.A. Third Edition. With 194 Illustrations. *In cloth gilt, price* 3s., *by post* 3s. 3d.

VAMPING. A Practical Guide to the Accompaniment of Songs by the Unskilled Musician. With Examples. *In paper, price* 9d., *by post* 10d.

VEGETABLE CULTURE FOR AMATEURS. Containing Concise Directions for the Cultivation of Vegetables in Small Gardens so as to insure Good Crops. With Lists of the Best Varieties of each Sort. By W. J. MAY. Illustrated. *In paper, price* 1s., *by post* 1s. 2d.

VENTRILOQUISM, PRACTICAL. A thoroughly reliable Guide to the Art of Voice Throwing and Vocal Mimicry, Vocal Instrumentation, Ventriloquial Figures, Entertaining, &c. By ROBERT GANTHONY. Numerous Illustrations. *In cloth, price* 2s. 6d., *by post* 2s. 9d.

VIOLINS (OLD) AND THEIR MAKERS: Including some References to those of Modern Times. By JAMES M. FLEMING. Illustrated with Facsimiles of Tickets, Sound-Holes, &c. Reprinted by Subscription. *In cloth, price* 6s. 6d., *by post* 6s. 10d.

VIOLIN SCHOOL, PRACTICAL, for Home Students. Instructions and Exercises in Violin Playing, for the use of Amateurs, Self-learners, Teachers, and others. With a supplement on "Easy Legato Studies for the Violin." By J. M. FLEMING. *Demy 4to, price* 9s. 6d., *by post* 10s. 4d. Without Supplement, *price* 7s. 6d., *by post* 8s. 1d.

WAR MEDALS AND DECORATIONS. A Manual for Collectors, with some account of Civil Rewards for Valour. Beautifully Illustrated. By D. HASTINGS IRWIN. *In cloth, price* 7s. 6d., *by post* 7s. 10d.

WHIPPET AND RACE-DOG, THE: How to Breed, Rear, Train, Race, and Exhibit the Whippet, the Management of Race Meetings, and Original Plans of Courses. By FREEMAN LLOYD. *In cloth gilt, price* 3s. 6d., *by post* 3s. 10d.

WIRE AND SHEET GAUGES OF THE WORLD. Compared and Compiled by C. A. B. PFEILSCHMIDT, of Sheffield. *In paper, price* 1s., *by post* 1s. 1d.

WOOD CARVING FOR AMATEURS. Full Instructions for producing all the different varieties of Carvings. 2nd Edition. Edited by D. DENNING. *Price* 1s., *by post* 1s. 2d.

www.ingramcontent.com/pod-product-compliance
Lightning Source LLC
Chambersburg PA
CBHW031411160426
43196CB00007B/976